CORDWOOD MASONRY HOUSES

CORDWOOD MASONRY HOUSES:

A Practical Guide for the Owner-Builder

ROBERT L. ROY

Sterling Publishing Co., Inc.   New York
Distributed in the U.K. by Blandford Press

For Ma and Pa Bates,
with love and fond memories of
our cordwood journey to New Brunswick

Styrofoam® is a trademark of Dow Chemical Company.

Copyright © 1980 by Robert L. Roy
Published by Sterling Publishing Co., Inc.
Two Park Avenue, New York, N.Y. 10016
Distributed in Australia by Oak Tree Press Co., Ltd.
P.O. Box K514 Haymarket, Sydney 2000, N.S.W.
Distributed in the United Kingdom by Blandford Press
Link House, West Street, Poole, Dorset BH15 1LL, England
Distributed in Canada by Oak Tree Press Ltd.
% Canadian Manda Group, P.O. Box 920, Station U
Toronto, Ontario, Canada M8Z 5P9
Manufactured in the United States of America
All rights reserved

Library of Congress Catalog Card No : 80-52325
Sterling ISBN 0-8069-5418-3 Trade
0-8069-5419-1 Library
0-8069-8944-0 Paper

CONTENTS

Color section follows page 64.

Preface

Cordwood is a tough habit to kick. It is intoxicating. It tends to free the imagination and change people's lives, and I have yet to observe a case where that change has been other than a happy one. So what is cordwood, some kind of drug? A mountain retreat?

No, cordwood construction is a method of building in which short logs are laid up in a wall like a rank of firewood. The wall's structural integrity comes from the special mortar matrix.

My wife, Jaki, and I decided to use cordwood masonry on our house after a week's experience in Arkansas helping to build a "traditional" cabin of heavy pine logs. We'd seen a couple of brief references to the cordwood technique in books and it struck us that we could handle short log-ends much more easily than those heavy fourteen-footers. Unfortunately, there was no practical literature on the subject, so we took a brief excursion into Canada's Ottawa Valley in search of old cordwood structures. We found two or three. On our way home to West Chazy, New York, we came across a man laying up a cordwood barn within a hefty post-and-beam framework. It looked simple enough. We decided that we were ready to go.

In 1975–76, we built Log End Cottage, a post-and-beam structure infilled with cordwood masonry. The story of the Cottage construction is recorded in my first book, *How to Build Log-End Houses* (Sterling, 1977) and is largely repeated in this book. Living in the house was an almost spiritual experience. Part of this feeling, I suppose, came from building our own home, but another part came from the warm, rustic atmosphere created by the cordwood within the framework of old, hand-hewn timbers.

Although we enjoyed the Cottage, there were two problems with living there. One was that it was too small for our growing family; the other was that it required seven full cords of wood to heat each year.

I was getting "building fever" again in 1977, and we decided to join the "underground" movement. Our new earth-sheltered home, Log End Cave, is considerably larger than the Cottage and is planned better with regard to living space — and we heat it on only three full cords per year. We still own the Cottage, which is only 120′ from the new house. As of this writing, it is occupied by friends who are building their own house nearby.

Having become incurably hooked on cordwood by our first building experience, we tried to incorporate the technique as much as possible in the

7

new home, and, as two years had passed, we were able to observe the effects of time on the various panels at the Cottage. We learned that the experimental sawdust panel was far and away the most successful in minimizing mortar shrinkage, and we used sawdust mortar exclusively at the Cave. The construction of this house is described in great detail in my second book *Underground Houses: How to Build a Low-Cost Home* (Sterling, 1979), and will be discussed later in this book.

One of the unexpected joys which has come out of our cordwood experience has been meeting and corresponding with other enthusiasts. Cordwood aficionados are a unique group of people: they are independent, uninhibited, and tend to laugh a lot.

The personification of the modern cordwood builder has to be my good friend and internationally renowned mortar stuffer, Jack Henstridge. His *Building the Cordwood Home* (Jack Henstridge, 1977), dealing with cordwood as a load-supporting curved wall, is probably the most widely read book on the subject to date. Jack has given tirelessly and unselfishly of his time in an effort to publicize the advantages of cordwood for the individual who wants to be free of the shackles of rent or mortgage. Jack has my admiration for the good work he has done and my gratitude for his generous help with this book.

Thanks also to all the cordwood people who contributed to Chapter Six, and to Dr. Allen Lansdown, of the Northern Housing Committee of the University of Manitoba, for permission to quote from the N.H.C.'s manual *Stackwall: How to Build It.*

Before we depart on our cordwood adventure, I should caution the reader that, like Strider in Tolkien's *Lord of the Rings,* cordwood is "known by many names." This already may have confused those who have seen brief references and articles in such magazines as *Farmstead, Harrowsmith,* and *The Mother Earth News.* To further complicate matters, three different books on the subject appeared at about the same time. Their titles were, *Building the Cordwood Home, How to Build Log-End Houses,* and *Stackwall: How to Build It.* "Cordwood," "log-ends," "stackwall," they all refer to the same thing. Add to the list, "stovewood masonry," "wood-masonry infilling," "stackwood," "firewood wall . . ."

Don't panic, we're all talking about the same thing: that type of construction where very short logs are stacked in a wall like firewood. In this book I will attempt to avoid confusion by using two basic terms. I will use *cordwood* as the general term for this type of construction, and *log-ends* as the name for the short blocks, butts, logs, ends, or pieces that are laid up in a cordwood wall. The reader should be aware of the other terms, however.

1. Overview of Cordwood Masonry

HISTORICAL PERSPECTIVE

In an article titled "Poor Man's Architecture" appearing in *Harrowsmith* #15, writer David Square says, "Curiously, the origin of the technique remains mysteriously obscure. In Siberia and in the northern areas of Greece, stackwall structures estimated to be 1,000 years old are still standing. Yet no one is certain where it all began."[1]

No accurate tracing of the cordwood masonry technique exists which comes near to determining a time or place of origin — it would be as easy, perhaps, to discover the family name of the first cave dweller — nor are the routes of cordwood's transmigration known with certainty, although a few relatively recent movements seem to be fairly well-documented. My own view is that cordwood masonry is such a simple idea that it may have been spontaneously conceived by many different people at different times and at different places. A few days ago I was riding along a country road in search of a missing dog. Attached to the front of a farmhouse I noticed a porch, the roof of which was supported by four sturdy posts. The people living there had stacked firewood between the house and the corner posts, and between the other posts, except for a clearway that had been maintained between the two central posts. I imagine that these woodpiles sheltered the front door from the winter winds very nicely — which may have been part of the intention — and provided close and easy access to the fuel supply. Given the situation, it would have taken little more imagination to see the possibilities of these cordwood walls as shelter.

Jack Henstridge conceives of the invention of cordwood construction in much the same way. As he tells it,

> The most precious thing that early man had was fire. He soon found out that unless he had a good supply of dry fuel the fire would go out, so he piled the wood up around a central fire. This worked great. Man not only had a good supply of fuel, but he soon found out that he could get in between the fire and the wood pile and keep warm. He then found out that if he built the pile

[1] Source notes appear on page 155.

high enough, he could lay long sticks over the top of it and some broad leaves over the sticks and the rain would not put the fire out. He also found that if he poked mud between the sticks in the pile, he could keep the wind out, too. All he had to do was to pile his circle of firewood so that the inside was even. He didn't have any saws or axes back then, so the outside was a veritable pin cushion. His enemies could not get in at him. Man had invented the "Cord-wood Wall" or more accurately the "Firewood Wall." He did not need the cave anymore.[2]

What I like about Jack's little story is its complete plausibility. In fact, cordwood masonry may have been discovered and rediscovered many times by events not unlike those described in Jack's fantasy. Jack further speculates that the Viking people brought the cordwood idea to North America about a thousand years ago. It is generally accepted that the Norsemen spent some time at the L'Anse de Mere Meadows on the northern tip of Newfoundland, where remains have been found that Jack likens to a round cordwood house, built with clay instead of mortar. The remains are too far gone to make this determination with certainty, but in discussing the matter Jack and I began to consider the following hypothesis: What if the Vikings built a round house with cordwood and plenty of clay and then, before roofing, built a huge bon-fire within the structure — or around its edges — and "vitrified" the walls, virtually welding wood and clay into one long-lasting monolithic wall. The ends of the cordwood would burn, but the fire would go out with the drawing away of the heat by the wet clay. A similar technique was used to vitrify earthen forts in ancient Britain.

The westerly migration of the technique was probably started by Scots from Nova Scotia who took the technique through Quebec, where it flourished, and into Canada's Ottawa Valley, where many old cordwood structures can be found. (They are not always easy to find, as people tended to plaster over the cordwood walls. Cordwood masonry has long been thought of as poor people's housing, and people like to cover up evidence of their poverty.)

Wisconsin became America's center for cordwood construction in the 1800's, and it seems that the technique's migration may have taken two different routes, one by way of western New York State and one from Canada into Michigan's Northern Peninsula and finally into Wisconsin's Green Bay area, where dozens of cordwood houses and barns were built in the late 1800's and early 1900's.

GENERAL REMARKS

Although the possibilities for individual initiative and imagination when building with cordwood seem to be unlimited, construction generally falls within one of three distinct categories, each of which will be discussed sepa-rately and at length. The purpose of this section is to give a general description of cordwood construction, and to compare some of the advantages, disadvantages and special considerations of the three distinct approaches to cordwood housing. This will give the reader a wider perspective when reading the chapters that are devoted to each individual style.

Before launching this discussion, two short but important notes might be appropriate.

1) I don't intend to favor any single method over the others. Rather than pursue a meaningless discussion over which method is "best" — there is no correct answer to this — I encourage people to choose the one that most appeals to them and best suits their individual circumstances . . . *and then to go out and build!*

2) It should be kept in mind that the three distinct construction techniques can be combined in various ways — there's no need to stick exclusively to one method. A house should reflect the owner's character and no two people are the same. I have seen about a hundred different cordwood homes — in person, in books and magazines, and through correspondence — and not only are no two alike, but no two are even remotely alike, except in their use of log-ends in the walls. Cordwood is like modelling clay: it just begs to be molded and caressed into a thousand different shapes. It is tempting to lecture about the economic and energy savings of cordwood, tell people how long these houses last, how beautiful they are, how easy to build, but the one concept which is hardest to impart to a prospective builder is that cordwood masonry is fun and good for the soul. Don't take my word for it . . .

The three different ways in which cordwood masonry can be employed in building are: (1) within a *post-and-beam framework,* (2) within a log or log-end framework of *built-up corners,* and (3) as a *load-supporting curved wall.*

POST-AND-BEAM FRAMEWORK

By this method the framework, not the log-ends, does the load-supporting. Our first house, Log End Cottage, is a good example of this style (see pages 120–121). I used to be almost dogmatic in my insistence that this method was the only way to go. No more! I now have equal respect for all three methods, as long as attention is paid to the special considerations of each. It may be that a well built post-and-beam framework is stronger per inch of wall thickness than the other methods, but it is difficult to build a wall as thick as is necessary for northern climates and still expose the posts and beams to view both inside and out. Jack Henstridge and I are now in agreement that 12″ of softwood (or 16″ of hardwood) is the absolute *minimum* insulation for Canada and the Northern States. Eight-by-eight posts will be recessed 2″ on each side of a 12″-thick cordwood wall — not as pleasing to the eye as hand-hewn posts and beams which are flush with the wood masonry infilling.

It should be emphasized that a 12″ to 16″ wall built by one of the other methods will be plenty strong enough, even for an earthen roof, *if built properly*. In southern areas, where such thick walls are not necessary, the post-and-beam method becomes extremely viable and can save on cordwood and mortar.

I must mention one final point in favor of post-and-beam construction: in conjunction with masonry infilling, the aesthetics of the house are outstanding. The Seven Dwarfs lived in a masonry-infilled, post-and-beam cottage. The diagonals add rigidity and beauty, but make the masonry work much more time-consuming. With log-end infilling, the diagonals aren't really necessary. The cordwood masonry will render the whole structure rigid.

BUILT-UP CORNERS

The built-up-corners method, also known as "stackwall," is an old style of construction and has something in common with each of the other methods. The cordwood is load-supporting (although many builders tie the corners together with top plates), but the technique is also that of masonry infilling, this time between laid-up corners of longer log-ends or 6-by-6 timbers, instead of between posts and beams. The advantage of the stackwall system is that a rectilinear structure can be built completely out of cordwood without the need for a post-and-beam framework. This is faster, easier, and usually cheaper than the post-and-beam method, though probably not quite so strong per inch of wall thickness. This slight disadvantage is more than made up for by the use of thick walls. The thick walls serve as a better insulation and thermal mass for those contemplating a cordwood structure in a northern climate. The finest example of this type of construction that I have seen is Malcolm Miller's incredible "cordwood castle" in Fredericton, New Brunswick (see pages C, 137–138).

Old stackwall barns are shown on pages 14 and 15, but their corners seem to be built up without mortar. Spiking the corners with 10″ spikes, such as are used in the building of horizontal log cabins today, would be one way of tying in these corners. Tom Kwiatkowski, building in Beekmantown, New York, built the corners of his twelve-sided home by nailing old 2-by-6's together (see pages F, 118–119).

CURVED WALLS

Thanks to Jack Henstridge, I have come "full circle" in my approach to cordwood masonry. The next cordwood house I build will be round, though

I still have great respect for the other two methods. If cordwood has taught me anything about life and personal philosophy, it is: "Don't be dogmatic." Who would have thought, for example, that a dome could be built entirely of cordwood masonry? Well, the staff of *The Mother Earth News* built one on their land near Hendersonville, North Carolina, and it appears to be structurally sound. While the structure is not a particularly outstanding example of the aesthetic potential of cordwood masonry, it does demonstrate the medium's tremendous inherent strength. If Jack wrote and told me that he was flying a cordwood airplane, I would reply immediately and ask him for his mortar mix.

However, there are a few undeniable advantages to building a round house that I want to mention. It is an indisputable function of plane geometry that any circle will enclose 27.3 percent more area than the most efficient rectilinear shape (a square) of the same perimeter. When compared to the more common rectilinear shapes that Western Man uses for his shelter, the gain of building round is more like 40 percent (see Fig. 1).

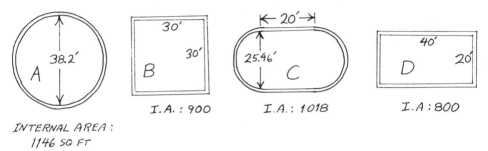

INTERNAL AREA :
1146 SQ FT

Fig. 1. The perimeters of all four of these house shapes are the same: 120 feet. The curved-wall shapes are a much more efficient use of cordwood.

Another advantage is that a round house will be easier to heat than a square or rectangular house with the same floor area. The economy of wall area mentioned above is part of the reason, of course, but the lesser wind resistance offered by a curved wall is also important. And if a radiant heat source, such as a woodstove, is placed at the center of the structure, there will be no cold corners, as all points on the circumference will be equidistant from the heat source.

And finally, as will be shown in Chapter 5, the circular house is the simplest to build.

CORDWOOD AS LOAD-SUPPORTING

Some of the principles of cordwood construction are different without the post-and-beam framework because we are asking the log-ends to support themselves as well as the roof load. But there is very little bond between wood and mortar. Are we asking for trouble? No, because we are really concerned

Stovewood Masonry

a "Barn near Crandon, Wisconsin.

an Indiana Grain Barn

WOOD BLOCK

Spring House in Montana

Stovewood masonry. (From An Age of Barns, by Eric Sloane, courtesy of T. Y. Crowell)

Fig. 2. Old stackwall barns.

with the compression strength of the wall, not the bond between wood and mortar. There are stone houses in the north of Scotland that have been occupied continuously for over 400 years and they have not a drop of mortar in the construction, no bond at all between the stones except the universal bonding agent: gravity.

The compression strength of a properly built round or curved-wall cordwood house is phenomenal. The strength comes from the mortar, which is used primarily as a levelling agent for the log-ends, not as a bonding agent. How can the wall compress? It supports itself all around the circle. (Means of stopping the wall from falling out are discussed in Chapter 5. It can't fall inward because of the *keystone principle*.)

THE MORTAR MIX

Here's where we come to the primary difference between cordwood as infilling and cordwood as load-bearing: the mortar mix. I have tried several different mixes for masonry infilling. Despite being protected every night by wet towels, all of the mortar without sawdust experienced shrinkage. I have found a mix of 3 sand, 4 sawdust, 1 Portland cement, 1 hydrated lime to be the best. The sawdust used is the heavy sawdust that a chainsaw or a ripsaw makes, not the fine stuff found where finished work is done, like in a cabinet-maker's shop. For a load-supporting wall, I strengthen the mix by leaving out one of the parts of sawdust: 3 sand, 3 sawdust, 1 Portland, 1 lime. (This is proportionally equivalent to a mix of 6 sand, 6 sawdust, 3 masonry cement, 1 lime.) A test panel that I constructed using this mix and old split-rail fencing for log-ends is absolutely monolithic — no cracks or spaces appear anywhere in the panel. The chart below gives several mixes, all of which have been successfully employed. This chart assumes equal parts by *volume*, not weight.

	SAND	SAWDUST	PORTLAND	MASONRY	LIME
N.H.C.[1]	6	0	2	0	1
N.H.C.[1]	5	0	0	2	1
Rob Roy[2]	3	4	1	0	1
Rob Roy[2]	6	8	0	3	1
Rob Roy[3]	3	3	1	0	1
Rob Roy[3]	6	6	0	3	1
Jack[4]	20	0	3	0	5
Jack[5]	12	4	0	6	3

[1] Northern Housing Committee of the University of Manitoba
[2] As infilling
[3] As load-bearing
[4] As described in *Building the Cordwood Home,* by Jack Henstridge
[5] Jack's sawdust mix

16

Jack's original mix is a bit "limier" than the rest, but if one of those parts of lime had been Portland, the resulting 20:4:4 would reduce to 5 sand, 1 Portland, 1 lime, very close to the others. In other words, independent research and experimentation have arrived at similar proportions: roughly 4 sand, 1 Portland, 1 lime. Other variations on these mixes will be noted in Chapter 6. The simplest of all is Cliff Shockey's mix of 3 sand, 1 masonry cement. Cliff, now building his second cordwood house, says this mortar has very little shrinkage. (Masonry cement, incidentally, is generally two parts Portland and one part lime.)

The sawdust in my mixes holds moisture for a week or two, which retards the set. By this method, mortar-shrinkage cracks are eliminated. Compared with other methods of retarding the set, such as frequently spraying the wall with mist or draping wet towels over the work, the sawdust method is vastly superior in that it is less trouble and retains moisture much longer. Dry log-ends — and they *should* be dry — tend to soak up the mortar's moisture. The sawdust resists this moisture loss from within. My experience is that the sawdust mix is just as strong as the others, stronger, really, because of the elimination of mortar cracks. At Log End Cave, I was unable to remove a log-end from a 10"-thick cordwood wall with a 9 lb. sledgehammer.

WOOD SHRINKAGE

' What little bond exists between wood and mortar is enhanced by reducing the shrinkage of both wood and mortar. The sawdust mix virtually eliminates mortar shrinkage. Wood shrinkage can be minimized by using only log-ends that have achieved what is known as *ambient moisture content,* which varies with different woods but is generally around 20 percent. Wood reaching this "steady state" moisture content will lose and absorb moisture in extreme weather conditions, but it will always return to about the 20 percent level. The mortar matrix, the source of strength in a load-bearing cordwood wall, can easily handle this slight working of the wood. What we try to avoid, by carefully drying the wood and mortar, is a condition of loose-fitting log-ends.

I have read in other cordwood literature the suggestion that the log-ends be soaked before they are laid up. The reasoning behind this suggestion is that the wet log-ends will not draw the moisture out of the mortar as rapidly. This technique is successfully employed by fireplace builders who, by soaking the firebrick, prevent an overly rapid drying of the fireclay. *I emphatically do not recommend this technique with cordwood!* Soaking the log-ends will swell them with water. True, this will assist in preventing the mortar from shrinking, but is not as effective as the sawdust mortar mix, as it does not address the problem of direct moisture loss to the air. The real problem comes when the swollen log-ends return to their ambient moisture content. The resulting wood shrinkage is unacceptable.

A very slight dampening with a whiskbroom of only the edge of the log-end will assist in promoting a better wood-to-mortar bond, especially with round cedar ends, which have a tendency to be slightly "hairy." But I repeat: *Do not soak!*

If all precautions fail and the wood still undergoes considerable shrinkage, the wall should be repointed inside and out after a year or two. This will be possible if the pointing is recessed as the construction progresses (which is advised in any case because of the more pleasing aesthetics of a wall so constructed). The wall is still strong, though it may be a little drafty without repointing if the log-ends have shrunk considerably. This advice should not be treated as a magic trick for building a house with green wood, which should not be done unless a temporary structure is desired. Besides, it would be a shame to lose the aesthetic advantage of showing log-ends *proud* of the background of mortar. Flush pointing is easier, to be sure, but does not accent the beauty of the wood nearly so well.

If only minor wood shrinkage is experienced after two years of living in the house, the tiny crack between wood and mortar can be filled by applying a coat of thick Thoroseal™ (or other cement-based masonry sealer) to the mortar joint with a ½″ brush. The use of white Thoroseal™ will brighten up the wall. This process is time-consuming, but only needs to be done once, if at all.

Incidentally, recessed pointing of ½″ or more is most easily accomplished in a wall constructed of split log-ends (page 24). It is easier to point this type of wall because it is possible to keep a mortar joint of a constant width by carefully selecting the pieces. This is impossible if you use round log-ends exclusively, because there will always be places where log-ends are tangent as well as those larger areas formed between three or more log-ends. It is much more difficult and time-consuming, therefore, to wiggle the pointing knife in and out and around the cylindrical log-ends. This is but one of the advantages of using split wood, which is discussed at greater length later in this chapter.

FINDING DRY LOG-ENDS

Green wood shrinks. If a cordwood wall is constructed of green wood, *even though a good non-shrinking sawdust mortar is employed,* the log-ends will be loose, the wall will turn drafty from the wood shrinkage, and the structural integrity will no doubt be diminished (although the compression strength of the wall might still be very high, as is evidenced by Jack's house). In short, the house will be unacceptable as a long-term dwelling unit. Chinking five thousand log-ends with caulking or oakum is a possibility and people have done it, but it isn't fun or pretty and it's a masochistic type of exercise which can and should be avoided.

How?

Use *dry* log-ends. Not almost dry. Dry.

If I belabor this point a little, it is because *wood shrinkage is the single most common cause of dissatisfaction with cordwood construction.* Before construction is begun, the log-ends should have achieved ambient moisture content, usually around 18 to 20 percent depending on the wood (and the ambient). The wood will expand and contract a little with the seasons and the internal heating input, but they will return to their original size. I have yet to hear of any cordwood builders having problems with log-end expansion in a wall. If there has been a problem with the working of the wood, the problem has always been one of wood contraction. So, if we want to use the driest wood possible, we have two options: finding dry cordwood or drying it on site prior to construction.

There are several potential sources of existing dry cordwood:

DEADWOOD IN THE FOREST. A large woodlot which has not been tended for several years may yield cords of dry deadwood in the form of fallen or leaning trees. If the bark is still on the wood, however, the chances are high that rot has already set in. Trees and dead branches without bark may well be dry and sound, however, as the first cut with a chainsaw will reveal. Admittedly, it would require a considerable woodlot and quite some labor time to supply enough cordwood by this method, but a couple of weeks of hard (but not unpleasant) work might make it possible to build that same season. In the northeast, elms that have fallen victim to Dutch Elm Disease might make a viable source of cordwood. Simply because the tree is dead, of course, does not mean that the wood is dry — it would have to be dead for several years for the slow drying process through side grain to have occurred.

Warning! Standing dead trees, even "stringy" elms, are known as "widow makers" in the logging industry, and with good reason. Large limbs can break off during the felling process, due to vibration. Hardhats, of course, would be mandatory during the felling of such trees, but the only completely safe course is to avoid them altogether in favor of wood which is already on the ground.

LOGGING SLASH. Another source of deadwood is to clear up the slash left after logging operations. Big loggers can't be bothered with short logs, tops, twisted pieces, and so on, and often leave them scattered on the lot. The debris that has managed to stay clear of the ground might make excellent cordwood after a few years. The owner of such a woodlot might be glad to have someone come in to tidy up the lot. Cordwood construction, then, can be beneficial to the environment. Jack Henstridge points out that by taking out the waste wood you give the other trees a better chance to grow. Jack also points out that fire-killed wood makes excellent log-ends; if you happen to be near an area that suffered a forest fire four or five years earlier, you may be spared considerable drying time.

OLD FIREWOOD. The kicker wall of Tom Kwiatkowski's home is built of old firewood. Perhaps a local firewood merchant has some split wood that is a couple of years old. Even at $100 per full cord, a "six cord house" could have its walls built for $600 plus the cost of the mortar. Remember that a cordwood masonry wall provides the support structure, the insulation, a phenomenal thermal mass, the exterior finish, and the interior decorating, so we can afford to spend a little extra on cordwood if necessary.

RECYCLED UTILITY POLES. I have heard of at least three houses made with recycled utility poles that had rotted at the ground but still had 20′ of good dry cedar or spruce just begging to be cut into cordwood. Cliff Shockey's two houses in Vanscoy, Saskatchewan, are made of old poles, and are super-insulated besides (see page 132). John Otvos, also building in Canada, paid $2 apiece for old poles and nothing at all for highway guard rails which were being replaced. Even when using this type of material, it is advisable to cut the log-ends to final size a month or more before construction, to give them a final air drying just in case any more checking occurs in the short pieces. This advice goes for any long pieces cut into shorter lengths, even the excellent material described in the following paragraph.

SPLIT RAIL FENCES. The best material for log-ends that I've found is old split-rail cedar fences. Not only are they dry — they probably achieved ambient moisture content fifty years ago — but their random shapes make it easy to lay up a wall with consistent mortar joints. Finally, if these old cedar log-ends are left protruding from — or proud of — the mortar matrix, the rustic character of the old rails is accented. The effect is that of a very old wall, except that it is in perfect condition and will stay so for a very long time. Now, I realize that these old fence rails are a rather specialized item and are sought after as decorative fencing in the suburbs, but I was still able to buy them in northern New York in 1979 for $25 per 16″ face cord, cut to length but not delivered.

The point is that log-ends are where you find them. Cliff Shockey found them on nearly treeless plains in the form of old utility poles. Even if you have to buy seasoned firewood for the job, it may well be a bargain.

But there are still two more options open to the owner-builder, both involving seasoning one's own cordwood.

SEASONING CORDWOOD FOR FUTURE USE

Economics and personal situations may dictate or allow the option of cutting all of one's own cordwood, barking it, bucking it to log-end length, splitting it (or not, depending on the owner's personal preference), and stacking it properly for use one to three years later. Let's look at this whole process now in much more detail, as it applies to all three styles of cordwood masonry.

KIND OF WOOD. First, we must deal with one of the most frequently asked questions concerning cordwood: What kind of wood should be used? In reply to this, I first have to admit that I have revised my thinking on this since the publication of *Log-End Houses*. I still believe that cedar is the best wood to use, if it's available, because of its high insulative value and resistance to decay, but I would now temper my previous advice which said: "If no cedar is available, use Douglas fir, western larch, or pine. Avoid white fir, hemlock, spruce, or hardwoods. They will rot." While the woods mentioned are grouped according to approximate insulative values and resistance to water-induced decay, it is nonetheless true that any non-punky wood can be used as long as basic precautions are taken, such as keeping the first course off the ground, protecting the wall with a generous overhang, and using fully seasoned wood. Remember, too, that the lime in the mortar helps to preserve the wood. Robert Johnson, of Rossville, Georgia, who has made the study of wood properties an important part of his life, wrote to me advising that white oak, locust, and sourwood are all highly resistant to rot, and that spruce and hemlock are fine if kept out of contact with the ground. I now believe that even white birch and poplar can be used if they are barked, split, and air-dried properly. Also, these and other woods that are known to have low rot resistance should not be included in the first three or four courses of the wall. Poplar, the ubiquitous "weed wood" that seems good for little else but pulp, has good resistance to heat loss when it is fully dry. Jack Henstridge says, "Poplar grows so fast that you can grow your own house in 10 years."

R-VALUE. More important than resistance to rot, which is most unlikely if the basic precautions are taken, is the wood's *R-value,* the measure of a material's ability to resist heat loss. The higher the R-value, the better the material's insulating ability. Softwoods — and I would group poplar with them, though it is not technically a softwood — generally have an R-value of about 1.25 per inch, with cedar at the top of the list. Here are comparative R-values for a few other building materials:

Material	R-value	R/inch
8" concrete block	1.00	0.13
12½" of stone masonry	1.00	0.08
4" common brick	0.80	0.20
1" hardwood board	0.91	0.91
6" fiberglass insulation	19.00	3.16

Working against us in cordwood construction is the undeniable fact that heat loss is greater along end grain than through side grain. I have even read where the heat loss along end grain is supposedly twice as great as through side grain, but tests performed by the University of Manitoba tend to disagree with this. There, a prototype was built with a 24" wall of poplar and spruce, with an insulated space between the inner and outer mortar joints. The walls of this structure were rated at R24, or about R1 per inch.

THERMAL MASS. It should be pointed out that the R-value of a material is only one part of determining the comfort level of a home. Spurred on by the manufacturers of insulation, great attention is paid to R-values while very little consideration is given to *thermal mass,* the ability of a substance to *store* heat. While 6" of fiberglass insulation is rated at R19, the thermal mass of the material is extremely low when compared with stone, which has a low

R-value but high heat-absorption characteristics. A massive house of stone will take time to heat because the stone "soaks up" the heat until it reaches an equilibrium where the loss is equal to the gain. Put the R19 insulation on the outside of the stone wall and the stored heat will be given back into the room. This is why underground house builders insist upon the insulation being on the outside of the concrete walls.

The need for building "a house within a house" to take advantage of both good insulation and mass heat storage ability is eliminated by the use of cordwood masonry, which does both jobs at once. The inner mortar joint of a cordwood wall is separated from the external mortar joint by an insulated layer, and acts, therefore, as a heat sink which gives its stored heat back to the room when the temperature drops. This helps maintain a steady temperature in the house over a longer period of time. The log-ends themselves also act as small heat sinks and help maintain steady temperatures, despite the fact that some of their heat is being robbed by the low temperatures outside. The external mortar joint is primarily structural, although there is probably some slight insulative value, especially if sawdust mortar is used. All in all, a cordwood wall has a phenomenal thermal mass, as well as a good insulative value, a combination unknown to any other single construction technique.

Incidentally, although hardwoods generally have a lower insulative value than softwoods, their thermal mass is greater. The trade-off point between these characteristics will not be established until a great deal more research is conducted. My gut reaction, based on my experience in the earth-sheltered housing field, is that American housing policy places too much emphasis on insulation.

To return briefly to the question of what type of wood to use: use what's available. This has been the guiding principle of "folk architecture" down through the ages: utilization of indigenous materials. It saves greatly on the energy cost of house construction, since transportation is greatly reduced. A fellow in Florida, for example, built his cordwood home of palm log-ends. People have written to me saying, "I haven't got any cedar . . . just oak. What am I going to do?" I reply, "Build of oak. And if you're concerned about the lesser R-value, build it 33 percent thicker than you would have with cedar."

BARKING THE WOOD

Jack Henstridge and I disagree with those who advocate the use of unbarked log-ends. We have several objections. First, the bark greatly inhibits the proper drying of the wood; second, insects love to get in between the bark and the epidermal layers of the wood; and third, although the rough bark might adhere to the mortar mix very well, as other writers in the field have pointed out, the eventual loosening of the bark-to-wood bond renders the bark-to-mortar bond superfluous.

Jack says,

An indication of the necessity of (barking) can be shown by using the town of St. Quentin, New Brunswick, as an example. In the early twenties the community was burned down and later rebuilt using the burned trees. The first houses built used relatively 'green' wood with the bark still on it. These houses did not last. The homes built later (approximately three to four years), after the wood had dried, shrunk, and the bark had fallen off are still standing today and as solid as the day they were built. . . . This was probably the last time (cordwood) construction was used on an extensive scale in North America.[3]

At any time of the year it is good to bark the tree soon after felling, and the same day if possible. The easiest time of the year to bark is spring, when the rising sap loosens the tight bond between bark and wood. In April, we could pull the bark off our cedar logs with our fingers. Sabre, our German shepherd, had a great time yanking 8' strips away from the log. That dog could really bark. The worst time of the year for barking is late autumn, when the sap has run back into the ground to protect the tree from freezing. The bark is then extremely reluctant to let go of the log.

As stated, we barked in the spring, so it was an easy job with an axe, a trowel, and a dog. We would force the blade of the axe or trowel under the bark at a tangent to the log, working the blade right and left. Because the wood was freshly cut, the bark pulled away like a banana peel. Later in the summer, Jaki and I helped friends bark cedar logs which had been cut a few weeks earlier. It was a completely different job. Armed with peeling spuds made from truck springs (a big improvement over a trowel and an axe), we were still lucky to complete one 14' log in an hour. The moral: peel fresh.

Though peeling logs was not a problem for us, it can be a troublesome task. Christian Bruyère's *In Harmony with Nature* (Sterling, 1975) includes an illustration by Robert Inwood showing clearly one method of making a handy peeling spud (Figure 3). Though I have not tried this tool myself, it makes more sense than the short-handled spuds we used on our friends' logs. The longer handle should give the user a vastly superior leverage.

Fig. 3. Making a peeling spud out of an old shovel.

CUT OFF &
SHARPEN
NEW EDGE

Tom Hodges, in a short article in *The Mother Earth News* (July, 1976), has another idea. He says,

> My debarker is simply a garden hoe with the blade straightened out. To make one just take an old hoe, heat its 'neck' until the metal is malleable, and bend the blade back until it forms a 165° angle with the handle. Sharpen the business end and presto! You've got a tool that's guaranteed to make easy work of any bark-stripping job. To use the spud, just anchor or wedge your (log) so it won't move, stand over it, and dig in. With a little practice, you'll soon be able to peel off two- to three-foot strips with one swipe.[4]

After the wood is barked, it should be cut into log-end size. A tractor saw (buzzsaw) is easier than a chainsaw for this job. Two different aesthetic effects can be obtained by splitting or not splitting the log-ends. There are examples of both styles in this book. The primary advantage of leaving the log-ends round is that fewer will be needed to build the wall, so there is less handling of materials. Also, some people just like the way they look.

SPLIT WOOD

Split cordwood has several advantages over the rounds. First, splitting the wood greatly accelerates the drying process. Drying of wood takes place primarily through end-grain and is extremely slow through the epidermal (outer) layers of the wood. Splitting the wood exposes more drying surface to the sun and air, especially in the heartwood area.

Second, split wood enables the builder to keep a constant mortar joint width, which facilitates pointing and strengthens the wall. Figure 4a shows rounds laid up in the most efficient pattern, a honeycomb configuration. You

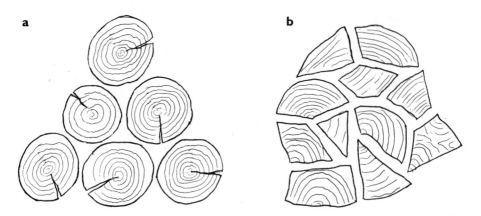

Fig. 4. A wall of split ends is easier to point.

can see that there are small mortar joints where the ends are tangent and a large triangular area of mortar between three adjacent ends. Figure 4b shows a section of split ends. The mortar joint is held constant, making it easier to use the pointing knife. Clearly, sample 4b is the stronger of the two. Try stacking split firewood in one rank and the same quantity of rounds in another, and then compare their weight-supporting abilities.

Another source of strength comes from the ease of pointing the randomly shaped split ends. The joint can be tightened by applying pressure with the pointing knife, more difficult to do with the different widths of the mortar joints in a round log-end panel. This pressure maximizes the bond between wood and mortar.

It is important to split the wood randomly — not all quarter logs, for example. A large selection of different sizes and shapes enables the builder to easily find the piece needed to continue the wall. Rounds and split ends can be combined in a wall, of course, and usually are to some degree.

DRYING

Whether or not the wood is split, it should be stacked for drying as early as possible. Log-ends won't dry if they are piled like stones. There are many theories about which direction firewood ranks should run for maximum drying. Some say north and south, others east and west. My advice is to choose a site out of the shadows and where there is good air movement. The top of a grassy knoll would be good. My own theory is that the line of the rank should be at right angles to the line of the prevailing wind, maximizing the wind effect in drying. The rank of wood should be off the ground, supported on logs or pallets. Figures 5 and 6 show different ways of supporting the end of a

Fig. 5. Pallets used to support cordwood for drying.

25

Fig. 6. Built-up corners support the end of this rank of wood.

rank of cordwood. Wood should be dried in single ranks for best air circulation. Finally, the top of the rank — but not the sides — should be covered to shed water. Split cordwood should dry at least a year, preferably two. Rounds might need as long as three years.

THE GREEN CORDWOOD HOUSE

There is yet another approach to the problem of obtaining suitable cordwood, and that is to build with unseasoned cordwood. This is the method which Jaki and I adopted at Log End Cottage and it worked very well for us. Late in the summer, after our post-and-beam framework was completed, we rented a portable circular saw and belatedly got around to cutting our barked cedar into the 9″ ends we had decided to use. Two friends helped. We set up an assembly line where Jaki would place the log on the movable table of the saw, I would cut the 9″ log-end, Joe Bernard would wheel them to the cottage, and his wife Pat would stack 'em up without mortar. In two days, all the panels were filled in with log-ends. We covered the exterior with cheap ½″ insulation board, stuffed the largest unfilled spaces with leftover fiberglass from the roof, and moved in in December.

The woodstoves dried the log-ends for a good six months before the weather and the garden allowed us to get back to work on the cottage. Considerable checking had appeared in the end grain (see Fig. 7). Beginning in June, we dismantled and rebuilt (with mortar) each panel, one at a time. A panel is a section between posts, usually about 5′ by 6′6″.

The drying method described above involves a lot of handling of materials. If, like us, you're trying to live in the house at the same time, tempers can run short as the living conditions become more and more chaotic. But it works.

Fig. 7. These log-ends were stacked to dry through a winter. The author points out a check that formed during the drying process. The panel on the left is still covered with insulation board. The panel on the right is complete.

THE TEMPORARY STRUCTURE

And now we come to one of the most exciting methods of obtaining cord-wood: the temporary structure. The temporary *structure* should not be confused with the temporary *shelter,* which is another good idea which we used to great advantage on our homestead. The temporary shelter is usually a small shed which you live in while building the main house. At our homestead, we built a 12′ by 16′ shed of plywood and economy-grade 2-by-4's for $350 and three weeks' work. We'd never built anything before, but a couple of friends showed us basic framing principles and we found that building one's own shelter isn't really all that difficult. We lived in that shed for eight months while building Log End Cottage. Today, there is a greenhouse attached and

Fig. 8. The temporary shelter, now a garden shed.

the building functions very well as a tool and materials store and a garden shed (Figure 8). In fact, I have temporarily moved back to the shed during the day to write this book, in order to get some peace and quiet. The place is well insulated against the 10° outside temperature and the old Quebec Heater ticking quietly in the corner is a cheerful reminder of our first year on the homestead.

But I'm romanticizing! The point is that with the temporary shelter, the homesteader will gain (1) a place to live free of rent or mortgage, (2) some building experience before tackling the "big one," and (3) a useful out-building for future use.

Now we will talk about the temporary cordwood *structure*. By this strategy the homesteader gains the first two above-mentioned advantages, but not the third. (Another equally important advantage is gained, however, as will be demonstrated.) This structure is truly temporary; it will be totally demolished in two or three years.

As has been stated, the single most common problem with cordwood construction occurs when people use wood that is not dry. Another common problem for folks trying to "get to the land" is economic: what with land payments, they haven't got the bucks to make the move, and in the meantime they are paying rent or mortgage and are nearing their goal at an economic pace inversely proportional to their shelter cost.

Enter the temporary cordwood structure. The land owner moves to the land even if it means tent living for a month (or perhaps the land is close enough to commute, although this loses the tremendous advantage of being on the site). He brings with him a good chainsaw (worth spending the money for this item), a trowel, a hammer, an axe, a wheelbarrow, a shovel, an old hoe, and lopping shears for easy tree trimming. Other necessary tools can be bought as needed, but the ones listed above will hold the homesteader in good stead for a long time, so he might as well get good ones. Tools are one of the primary expenses for the temporary cordwood structure, but they will be used again on the permanent dwelling and for other jobs, so it's money well spent.

Here's the basic idea: the owner-builder goes out on his woodlot, cuts cord-wood like a madman, mortars it up on the cheapest temporary foundation he can think of, and roofs it with 1-by planking and 30 lb roll roofing (or some other low-cost roofing). He moves into the structure and heats it with a woodstove. (If the economics allow, he might as well get a good one that can be used again when the house is rebuilt.) When the cordwood is dry, the owner/builder demolishes the structure and rebuilds it on a proper foundation which he has built in the meantime.

At this point I can imagine a certain part of my readership heading for the exits. But wait! Come on back and waddle through the next chapter with the rest of us. It might be fun, and there are a few useful tips that apply to all styles of cordwood construction. And remember: you don't *have* to be crazy to build of cordwood, although it's no great disadvantage if you happen to have been blessed with this vantage point. And now for the details . . .

2. Construction

FOUNDATION

If the temporary structure is to be built in a cleared area, the ditch-and-crushed-stone foundation would serve as well as any. A flat stretch of land should be chosen, or created with a bulldozer. A ditch 1' deep and 2' wide should be dug for cordwood walls between 12" and 16" wide, proportionally larger for wider walls. This ditch should be dug by hand where conditions are favorable. If it is necessary to install an approved septic system, this foundation ditch could be dug at the same time with a backhoe.

The cheapest, easiest temporary cordwood structure would be built using the curved-wall method described in Chapter 5. The shape of the permanent house is in no way restricted by the shape of the temporary structure, but it is imperative that enough wood be used in the first to build the second.

Estimating cordwood required is a simple matter and, as it is common to all three methods of construction, this seems as good a time as any to explain the calculations involved. The most convenient measurement of cordwood is . . . *cords!* Here, it is of paramount importance to differentiate between a *true cord,* which measures 4' high by 4' wide by 8' long, and a *face cord* (also called a *run*) which is 4' high by 8' long by whatever length of log the merchant is selling: foot-length, 16", 18", whatever. This can get very confusing, so we will stick to full or true cords in our calculations.

Incidentally, as a true cord has between three and four times as much wood as a run of "foot-wood," always make sure of a dealer's definition of "cord" before buying wood for any purpose. A true cord is not exactly four times more voluminous than a 12" face cord because the foot-length stuff packs much tighter than the more irregular 4' lengths. A full cord cut into 12" log-ends and restacked may yield only three to three-and-a-half face cords, depending on the straightness of the original logs (Figure 9).

This compression of the piles in the restacking can be incorporated into our calculations very nicely, as the loss of wood will be more than made up by the gain of wall area through the mortar joints in a cordwood wall. So, if we deal in 4' cords — a convenient size for handling in the woods — we can disregard both the mortar area of the finished wall and the disappearance of wood when it is cut short and restacked. Nice. And there should be wood to spare, which is a whole lot better than being short of seasoned log-ends.

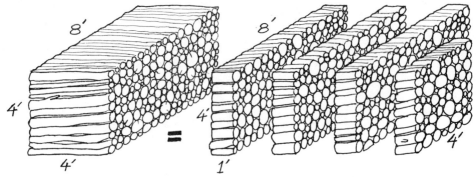

Fig. 9. The strange case of the disappearing cordwood!

One good thing about cordwood masonry is that the leftover material makes darned good firewood.

The chart shown can be used to determine the number of full cords of wood required to build an 8'-high wall with an inside perimeter of 120' at various wall widths. The wall widths given have been chosen because log-ends at these lengths can be taken from 4' cordwood with no waste. Because ¼" is lost with each cut of the saw, the average "real" wall width is slightly less than the full widths given. Care should be taken that the wood is hauled out of the forest in lengths equally divisible by the length of log-end required. Otherwise, you run into problems of wastage and unnecessary sawcuts. It is assumed that wood "loss" due to tighter stacking at the shorter length is equalled by wall area gain because of the mortar joint.

A	B	C	D	E	F	G
			Area of wall			
	Gross in-		covered		Adjusted	Adjusted
	ternal wall	Face	by full	Full	wall	full cords
Width	area	cords/	cord	cords	area	needed
of	(120x8)	full cord	(Cx32)	needed	(.85B)*	(F/D) or
wall	(sq ft)	(48"/A)	(sq ft)	(B/D)	(sq ft)	(.85E)
6"	960	8	256	3.75	800	3.125
8"	960	6	192	5	800	4.167
9½"	960	5	160	6	800	5.0
12"	960	4	128	7.5	800	6.25
16"	960	3	96	10	800	8.333
24"	960	2	64	15	800	12.5

* Assumes 15 percent of wall devoted to windows, doors and framing.

We will keep our examples simple, as I cannot possibly entertain the infinite number of shapes that can be built with cordwood masonry. Note that the measurements given are *inside* measurements. On rectilinear structures, the corners will be either post-and-beam or some sort of built-up square timbers. Either way, they do not enter into the cordwood calculations. With a curved wall, the mortar joints are wider on the exterior than on the interior, but only fractionally. Wider mortar joints are okay, but narrow ones are unacceptable when cordwood is load-supporting. Also, the use of interior measurements means that we are calculating *actual usable internal square feet*. With the thick walls of cordwood construction, usable square footage is much more meaningful than gross square footage calculated from external dimensions. Even though the floor plan of the permanent house may not yet be known, it is good to know approximately what size house can be built with the amount of cordwood you seasoned in your temporary structure.

Four sample house-shapes, with internal dimensions and areas given, are shown on page 13. They are all drawn to the same scale and all have interior perimeters of 120'. The statistical analysis on page 30, then, holds true for all four examples of shape, and assumes an 8' wall height. Gable ends are not considered, but should be added if necessary.

It should be noted that in Figure 1, the round house, *A*, has 1146 usable sq ft, compared with 900 sq ft for the square house, *B*. (These shapes are the most efficient for curved walls and rectilinear structures, respectively.) Expressed as a percentage, *A* has 27.3 percent more usable floor area than *B* for the same amount of cordwood. Curiously, *C* also encloses 27.3 percent more area than *D*, again showing the advantage of the curved wall. Numerologists may find it interesting that if a refined value of pi is used in the calculations, the area of the rectilinear part of the house shape, *C*, is exactly equal to the sum of the areas of the two semicircles (509.296 sq ft), a very interesting shape, indeed, and probably the most practical for a temporary cordwood structure. No external wall posts, beams, or built-up corners would be required, as in *B* or *D*, and the rafter spans are reasonable to work with (unlike *A* which would require tremendous rafters or an internal post-and beam framework as described in Chapter 5). Another advantage of *C* is that it can be elongated to virtually any length, in order to season greater amounts of cordwood. (Ferny Richard's cabin utilizes a similar shape — see pages B and 130.)

Another way of seasoning more cordwood would be to build 10'-high walls in the temporary structure, whereas only 8' walls would be used in the permanent home. This effectively increases the cordwood amount by more than 25 percent, as the added 2' of height would contain no windows or doors.

A better idea might to be build an internal cordwood wall as shown in Fig. 10. If 16" wood is used, the resulting wall would dry two full cords more than the 8.33 required by the above chart. The percentage gain would be the same with other log-end lengths. If corners such as shown in Figure 75 on page 82 are used at the wall intersections, there is the advantage that half of the temporary structure can be dismantled and rebuilt in the permanent house, while you are still living in the *module* created by the dividing wall. Then, the owner/builder can move into a completed module of the new house. Shelter is thus provided during the remaining dismantling and rebuilding. This plan would work if the second home had as a part of its design a section that could be completed and roofed independently of the rest.

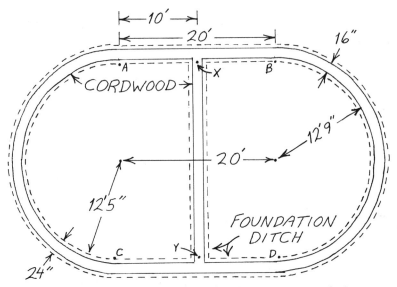

Fig. 10. Foundation ditch plan for the temporary shelter.

THE DITCH-AND-CRUSHED-STONE FOUNDATION

A flat site has been chosen and it is time to put in the foundation for the temporary structure. We will assume a 16″-wide cordwood wall, requiring a 24″-wide foundation ditch. Two round stakes are driven in the ground 20′ apart. Using a mason's line with a large nail at one end and a loop at the other, and with the stakes as center-points, two large semicircular arcs, each with a radius of 12′5″, are scribed on the ground. Lines *AB* and *CD* are drawn tangent to the curves created. In theory, these straight segments should each be exactly 20′ long, and the diagonals *BC* and *AD* should be equidistant. Now, with the scribed earth as the inner bounds of the ditch, a trench 2′

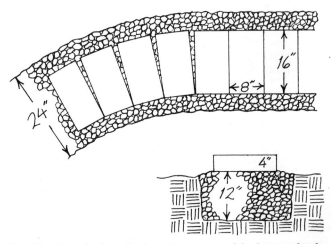

Fig. 11. Foundation ditch and concrete blocks method.

wide by 1' deep is dug all around the perimeter, and a similar ditch is made to join midpoints *X* and *Y*. If digging is difficult with "idiot sticks" (picks and shovels), get a backhoe on the job. If the backhoe is coming to dig for the septic tank and leachfields anyway, it will cost little extra to do this ditch. It will be necessary to tidy up the machine work by hand, however, as backhoes don't dig around curves so well. Theoretically, 11 cu yd of crushed stone should fill the ditch. I'd order 12; 14 if a machine was used to dig. The stones should be tamped.

Now, there are two alternatives. The cheap way is to start laying cordwood directly on the ditch of crushed stone. A better way would be to lay out 217 4" by 8" by 16" solid concrete blocks as shown in Figure 11. When the house is dismantled, the blocks will be salvageable for some other purpose. They should be vibrated in with a back-and-forth sliding motion so that they are well supported by the crushed stone. They don't have to be absolutely level with one another. That's what makes this type of building progress so quickly: fine finished workmanship is not as important as with a permanent structure.

THE INTERNAL SUPPORT STRUCTURE

In the suggested plan three support posts are recommended, one at each end of a 20' ridgepole, and one supporting halfway. The posts should be straight, roughly 6" in diameter and 16' long. If smooth, round posts are used, they will come in handy when it comes time to lay up the cordwood, as will be seen. The posts should be set a full 3' in the ground and "jammed" in the hole with large stones. Tapping these jammers with a sledge will allow for plumbing of the post. A plumb bob or level will serve to check the post for plumb (Figure 12).

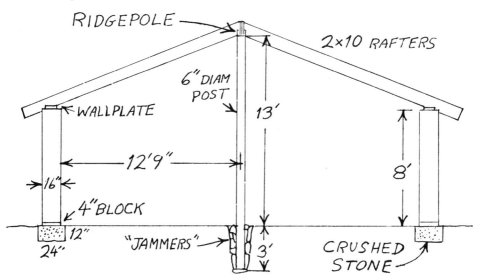

Fig. 12. Temporary structure dimensions.

Thirteen feet of each post will be above ground. It is important that the tops of the three posts are straight and level with one another. The ridgepole is constructed of 2-by-10's and 2-by-4's, as shown in Figure 13. The 2-by-4's offer a shelf upon which the rafters can rest. If the ridgepole is too heavy to install in one piece, it can be built in place.

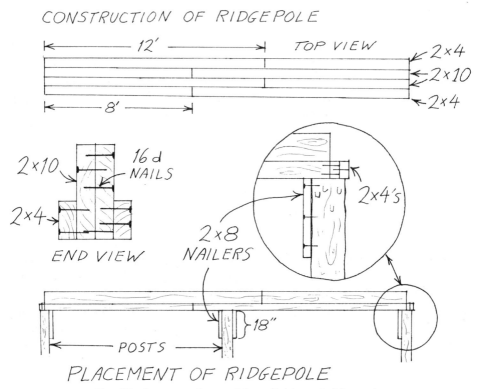

Fig. 13. Construction and installation of ridgepole.

It is wiser to buy full-sized, rough-cut lumber from a sawmill than the finished stuff offered at building supply yards. The rough-cut is a lot cheaper because it is local, hasn't been planed, and is handled by fewer people. It is stronger because it is full-sized. For example, a rough-cut 2-by-4 has 8 sq in of cross-sectional area, whereas a finished 2-by-4 measures 3½″ by 1½″, only 5¼ sq in of cross-sectional area. All else being equal, the rough-cut is about 50 percent stronger! The problems with rough-cut are that (1) it is difficult to use with standard framing practices because of size irregularities, and (2) it is often freshly cut and is, therefore, green. These are not problems in the temporary structure. In fact, the strategy for drying cordwood works equally well for seasoning the rough-cut. It will be properly dry when used again on the permanent home. Finally, cordwood masonry is primarily a roughhewn, rustic kind of construction which lends itself more to the use of rough, rather than finished materials.

HANDLING THE WOOD

It would be great, of course, if the 10.33 cords required to build this house were already cut to length, split, and stacked on site. Even a month's drying is a whole lot better than none at all.

The wood is cut to 4′ lengths for easy handling. Near the site, it is "bucked" into 16″ lengths, either with a chainsaw or with a rented circular saw such as farmers attach to their tractors with a belt. The wood should be split, but barking is optional. If it comes off easily in the 4′ lengths, fine; otherwise, wait until the wood has properly seasoned in the temporary structure. If construction is to begin immediately, dry stacking to measure cords is really a waste of time. When the structure is complete, the right amount will have been stacked!

The technique for laying cordwood is similar to that described in Chapter 3 of this book, with three differences. First, a weaker mortar mix is used: 6 parts sand to 1 part masonry cement. This mix is strong enough to support the wall, but will break loose of the cordwood easily when it comes time to demolish the wall. Second, insulating the mortar joint seems hardly worth the bother if only two winters are to be spent in the structure. I'd recommend two 4″ mortar joints with an 8″ dead air space between. The temporary structure will not be as easy to heat as the finished home, especially towards the end of the second winter when air will be leaking through the checks and gaps around the log-ends. But this will encourage you to keep the fire stoked, which, in turn, will dry out the log-ends better, the very point of this endeavor. A couple of years of minor hardship is a small price to pay for the end result: a beautiful, permanent, energy-efficient home that is paid for! Listen to this:

We spent our first winter in upstate New York in a cordwood home with no mortar in the 9″-thick walls. The worst night was when the neighbor's cat, for whom we were caring, decided to use our sleeping bag for a litter box. We washed the bag and hung it up to dry over an internal beam so that the water would drip into the kitchen sink, 6′ from an old cookstove that I kept going all that night. I also kept an all-night fire in a good parlor stove in the same open-plan room. Jaki and I retired to the loft and tried to squeeze ourselves into an old mummy-type, single-passenger sleeping bag. The temperature that night was −30° F., with 40 m.p.h. winds creating a windchill factor of −100° F. Snow was blowing through the wall and accumulating inside. The wet sleeping bag over the sink was frozen solid by morning, despite the woodstoves. We survived and have a great "cold story" to relate when the homesteaders start bragging about that awful winter of so-and-so. I'd like to report that the story had a completely happy ending featuring Curried Kitten, but our neighbors returned in time to rescue the little beast.

The third difference in the laying-up technique is that neat and tidy pointing is not needed. Only rough pointing is necessary, and this just to save on mortar; no sense letting it come glopping to the ground.

DOORS AND WINDOWS

The doors should be framed by one of the methods described elsewhere in the book (see page 78). Location of windows is not so critical as it would be in a permanent house. When it looks like it's time for a window, put one in. So what if it's 36" off the floor instead of 39"? Cheap recycled doors and windows are fine. Windows can be double-glazed for the winter with 6-mil clear plastic.

RAFTERS

Plates should be installed at the 8' level as described in Chapter 4. Rafters should be 2-by-10's birdsmouthed at each end to rest properly on the plate at their lower ends and the little support shelf formed by the 2-by-4's on the ridgepole. Centers are 24" on the rectangular part of the structure and vary on the semicircular sections, as shown in Figure 14.

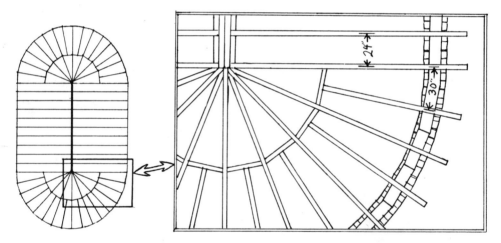

Fig. 14. Rafter plan for the temporary structure.

ROOFING

Planking can be recycled boards or rough-cut 1-by material. It's a good idea to make the new rough-cut boards uniform by passing them once through a planer. This makes for easier roofing and is particularly advisable if the material is to be used again on the permanent house (as it should be), because no sawmill will allow recycled material through their planer. There's too much chance of a nail slipping through and damaging equipment.

The cheapest possible temporary roof covering should be used. Fifteen-pound roofing felt is just too light, however. I recommend 30 lb roll roofing.

It will not be feasible to recycle this material. The crushed stone, the mortar, and the roof covering are probably the only total "write-offs" in the temporary structure.

INSULATION

This should be 23″ by 6″ fiberglass, which is why 24″ rafter centers are recommended. Clearly, insulating becomes tricky on the semicircles; lots of angle cutting is necessary. If the material is to be recycled by the use of a similar rafter system, there are no problems. And, too, fiberglass of unusual shapes can always be cut into strips and recycled as wall insulation when the permanent cordwood walls are built.

FLOOR

There are several possibilities for a floor. The easiest, most natural is an earthen floor. Or, if that's just too rough, a weak unreinforced 2″-thick concrete floor could be poured by a ready-mix truck. This would be a "write-off," but should be fairly easy to break apart with a sledge for removal to a land fill.

Then there's the pallets-and-rough-boarding method. It is often possible to obtain industrial pallets, such as blocks are delivered on, for nothing or next to nothing. Broken pallets and crates or new rough-cut boarding could be nailed to a framework of carefully placed pallets.

Well, that's about it. Add a pitcher pump, a simple gray water drain, an outhouse, and a stove or two, and the house is ready. I recommend a Metal-bestos chimney, which is expensive but of good quality. And it can be used again in the permanent house.

Now, admittedly, this strategy is not for everyone, but there are many ways by which the plan can be altered to suit those of greater or lesser financial means. Imagination is the limit. The advantages are worth repeating: the ability to move to the land immediately, gaining valuable building experience, seasoning the cordwood, eliminating the current rent or mortgage situation, and the extra time to find materials and to lay a proper foundation for the permanent home. This last point cannot be overemphasized. By the "pay-as-you-go" method on the new home — made possible by the low cost of living in the temporary structure and because of the wealth of materials and tools already paid for — *it will be possible to avoid the mortgage road altogether.* The requirements to play this game are: land (preferably with woodlot), a relatively small amount of cash (depending on personal fiscal circumstances and attachment to creature comforts, such as floors — see chart), and eight weeks of work (or 16 part-time weeks) for two inexperienced people. And courage.

Temporary Structure Costing Estimate

(based on 1980 prices)

Expense	As outlined in text	Survival budget
Chainsaw	$ 300 (new)	$ 150 (used)
Other tools	300 (new)	100 (used)
Ditch and drain	40* (backhoe)	0 (backache)
Crushed stone	70*	70*
Blocks	90	0 (none; or flat stones)
Cordwood from land	30 (gas, oil, saw upkeep)	30
Sand	80*	80*
Masonry cement	120*	120*
Windows and doors	100	40 (the true scrounger finds a better deal)
Posts	0 (cut on land)	0
Ridgepole, rafters	400 (rough cut)	20* (cut from the land, saw expenses)
Roof planking	600 (rough cut)	20* (recycled; hauling)
30-lb. roofing	80*	80*
Insulation	300	300 (expensive; sometimes this can be recycled)
Miscellaneous	80*	40*
Outhouse	100*	50*
Stove	500	50* (cheap stove; or bought at auction)
Stovepipe	200 (Metalbestos)	50* (homemade system)
Pump and pipe	50	50
Total investment	$3,440	$1,250
* Write-offs	570	580
Re-usable materials	$2,870	$ 670

* Write-offs are materials that cannot be reused.

The disadvantages are the extra couple of months of work involved in providing one's own permanent shelter, and, of course, the hardship involved. Now, these aren't even disadvantages if one believes, as I do, that it is necessary to build two houses to get one right, and that a certain measure of hardship is good for the soul. If one has a mind to "get back to the land" and to be self-sufficient, well, the plan described above certainly throws the novice into the thick of things . . . a sort of on-the-job training.

3. Cordwood Masonry Within a Post-and-Beam Framework

This chapter is adapted from *How to Build Log-End Houses*, but has been rewritten and updated in light of new information that has surfaced in the field during the past four years. In the first book I was quite insistent that cordwood within a post-and-beam framework was the only safe way to go. The structures described in Chapter 6 and the living proof of century-old buildings have convinced me that my original fears concerning the strength of cordwood masonry as load-supporting can be disregarded if care and elementary caution are exercised during construction.

FOUNDATION

At Log End Cottage, we have a full cellar and a block-and-mortar foundation. The blocks are 8″ wide and are concrete, not "cinder blocks" (which are cheaper but not as strong). A poured concrete foundation would be good, but much more expensive because you'd have to call in a contractor. If I were building the Cottage again, I would use a surface-bonded, block-wall foundation.

Surface bonding is the application of a special cement-based material to both the exterior and interior surfaces of a wall of concrete blocks laid up without mortar after the first course. The first course is mortared solely to establish a level base for the further stacking of the blocks. No mortar is ever used at the end of blocks, even on the first course. The secret of surface bonding — made by Conproco, Surewall, and others — is the thousands of tiny glass fibers which permeate the mix. When applied correctly, to a thickness of ⅛″ to both surfaces of a dry-stacked wall, surface bonding has been said to be six times stronger against wall flexure than a conventionally mortared wall.[5]

My initial skepticism of surface bonding was dispelled when I used it at Log End Cave and began to get a sense of what I was actually doing to the wall. The only way that the wall can fail is if the ⅛″ membrane were to rupture along its entire length, but this is prevented by the tens of thousands of short glass fibers bridging the tiny spaces between adjacent blocks. This method of construction has something in common with screwing and gluing plywood to both sides of a 2-by-4 frame: the resulting panel has a strength which seems to exceed the inherent properties of its components.

Most people — and I used to be one of them — think that mortar "glues" blocks together. This is not strictly so. Mortar's primary purpose is as a leveling agent. When a block wall fails, the cracks in the wall invariably follow the mortar joints. Rarely do the blocks break. People say to me, "It would seem that the strongest block wall would be mortared *and* surface-bonded." I think not. The mortar will still be the weak link in the chain, and the surface bonding will be of little help because the mortar will diminish the ability of the individual ½″ glass fibers to span the gap between one block and another.

Building a surface-bonded wall is not difficult if the builder pays attention to detail and proceeds carefully. The footing should be as near to level as possible. We shuttered and poured our own footing at the underground house, ably assisted by a good friend who was in the contracting business. Frequent use of a surveyor's level (Figure 15) will assure that the footing is level within ¼″, but the first course of blocks must be even closer than that for dry stacking, thus the need for leveling the first course in a bed of strong mortar. Again, a surveyor's level and a measuring rod are the right tools for the job. The use of a 2′ level along the width of the block will keep the wall plumb.

A friend and I stacked 900 blocks in five days, being extremely careful to keep our courses plumb and level. Minor leveling of blocks is accomplished with wall ties or shims cut from aluminum printing plates. Wooden shims should not be used. It is important, too, to keep the ends of each block butted against one another. Figures 16, 17, and 18 show surface bonding at various stages of construction.

It is beyond the scope of this book to give a detailed step-by-step discussion of the surface bonding technique, or to outline its various special considerations, but the information is readily available in a pamphlet called *Construction with Surface Bonding,* printed at the U.S. Government Printing Office (see Source Notes). Also, I discuss the technique thoroughly in my book *Underground Houses: How to Build a Low-Cost Home.*

THE FOOTING

The rule is that a footing's height should be equal to the width of the wall, and its breadth should be twice the width of the wall.

Fig. 15. Levelling the footing forms.

Figs. 16, 17. (Left) The first course of blocks is set in a bed of strong mortar. (Right) Corners are built first, then a mason's line is stretched to guide the placement of the other blocks.

Fig. 18. Applying the surface bonding mix with a trowel.

41

BACKFILLING

At the Cottage I made the mistake of backfilling with the same material that came out of the hole, and which had poor drainage characteristics. Had I backfilled with sand or gravel, allowing water to percolate down to the footing drains, I might not have experienced the ensuing wall failure. At the end of the second winter of occupation, the top three courses of blocks moved in about $\frac{1}{16}''$ because of hydrostatic and frost pressures on the wall. This was enough to break the Thoroseal™ waterproofing and we ended up with an inch of water on the basement floor. We installed roof gutters for the third winter and that reduced the hydrostatic pressure on the wall sufficiently to cut way down on the water problem. We braced the wall from the inside with strong wooden buttresses and there has been no further wall movement. A surface-bonded wall backfilled with sand would have prevented the problem, as we have proven at Log End Cave.

FRAMEWORK

Builders in our area look at the framework of Log End Cottage (Figure 19) and shake their heads in disbelief. They say they've never seen a frame built so strong. Some call our house "overbuilt." Is a house overbuilt because it lasts 300 years instead of a single lifetime?

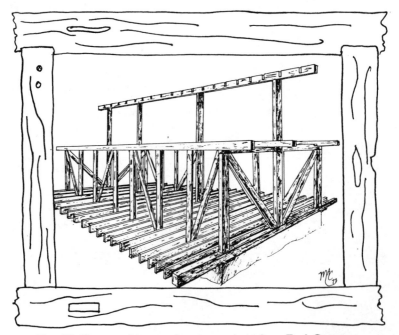

Fig. 19. Post-and-beam framework at Log End Cottage.

The post-and-beam framework should do all the load supporting. The log-ends, in this method, serve as masonry infilling. Masonry infilling was common in Elizabethan times and is found in old houses all over Europe. Many of these houses are still in use after hundreds of years. The advantage of log-ends, as opposed to the lath and plaster used in England, is their insulative value. If the mortar joints are also insulated — a must in the north — the wall has an "R" value of between .9 and 1.25 per inch of thickness, depending on the kind of wood used.

At Log End Cottage, the largest span between posts is 8', measured on center. This situation occurs at both gable ends, so we chose our largest and best pieces as corner and gable support posts. Along the side walls, no span is greater than 6' on center. Our original intention was to add a sod roof in the future and we wanted to be sure that the framework would support the tremendous weight. The roof finally ended up with cedar shingles, so we really didn't need to build the frame anywhere near as strong as we did.

Besides corners, posts are needed to frame doorways and large windows. Small windows can be "floated" in a cordwood wall as shown in Figures 142 and 143 on page 149.

Old barn timbers, if available, are ideal companions for log-ends. They are dry and, provided care is exercised in their selection, incredibly strong. Avoid punky or insect-infested beams. Many old timbers have had woodworm in the past, but the worms have long since abandoned the wood after it became too dry for them. No need to fear these timbers, but make doubly sure the worms are gone by excavating a little with a knife. Judging the structural quality of old beams is an acquired skill, but play it safe while acquiring it: if in doubt about a beam, don't use it. Beams that aren't good enough for construction make excellent borders for raised-bed gardens.

If old barn beams are not available, I would advise building box posts and beams out of 2-by-10's and 2-by-6's, as shown in Figure 20, before I'd suggest cutting 10-by-10's from one's own logs. Again, get dry material, even if

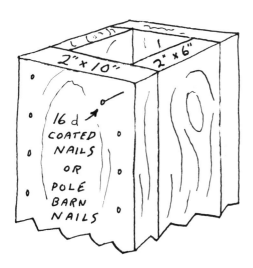

Fig. 20. A box post.

it means horse-trading with a sawyer as I did, trading green wood for a lesser amount of dry lumber. With cordwood walls, green posts will shrink away from the masonry, so use barn beams or dry box beams. Or use new posts and beams and be prepared to run a bead of latex caulking around the framework when it shrinks. I have seen this done successfully, and it doesn't look too bad.

Figures 21, 22, 23, and 24 show four views of the framework. After these photographs were taken, we added diagonals along the side walls just as we did at the gable ends. The diagonals are paired 3-by-10's with a 3″ insulated space in between. They are, therefore, exposed both inside and outside. Diagonals add rigidity to a post-and-beam structure because they form triangles. In modern framing, plywood serves the same purpose. My thinking now is that the diagonals, even though they are very attractive, are more trouble than they are worth. The cordwood masonry will supply all the rigidity required. However, temporary diagonals should be employed until at least one panel of cordwood infilling is completed on each wall.

As a point of interest, our floor joists and roof rafters are also old 3-by-10's, 2′ on center. We were lucky to find excellent buys on 3-by-10's from people who were demolishing an old school and an old hotel not far from our homestead. See Figures 25–27 on pages 46–47.

All posts should be checked or toenailed into a wooden plate fastened to the masonry foundation. Fastening the plate to the foundation is discussed on page 47. It is imperative that the top ends of all the posts are tied together

Figs. 21, 22. (Left) Post-and-beam framework. The interior beams at the far end support the sleeping loft. (Right) Post-and-beam framework, north gable end.

Fig. 23.
Post-and-beam
framework, south
gable end.

by the use of another plate. In Log End Cottage we used heavy barn timbers, as much for aesthetics as anything else. Doubled 2-by-8's or 2-by-10's would certainly accomplish the same thing. With the tops of the posts firmly tied together by a plate, the pressure exerted by the masonry panels is resisted.

At Log End Cottage we used two large barn beams to tie the long east wall to the west wall. These beams fly from plate to plate through the kitchen/dining area — which has a cathedral ceiling — and they make a convenient

Fig. 24. The posts of the west wall.
Later, a porch was added, using
the cantilevered three-by-tens for
support.

45

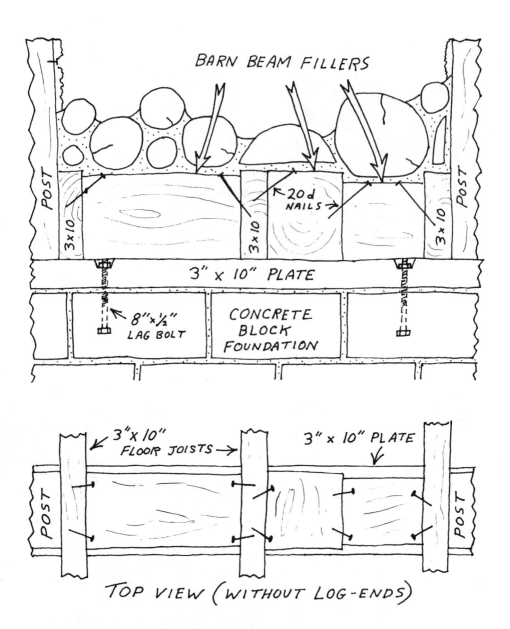

BARN BEAM FILLERS

POST

3×10

3×10

←20d NAILS →

3×10

POST

3" × 10" PLATE

←8"×½" LAG BOLT

CONCRETE BLOCK FOUNDATION

POST

←3"×10" FLOOR JOISTS →

3" × 10" PLATE

POST

TOP VIEW (WITHOUT LOG-ENDS)

Fig. 25. In Log End Cottage, the author used three-by-ten floor joists, 24 inches on center. The joists cantilever out over the foundation plates (also three-by-tens), supporting the front porch and the back firewood deck. It is convenient to fill in the space between joists with segments of barn beams left over from the framework. The barn beam fillers add support to the joists and establish a fairly flat surface upon which to be laying log-ends. The barn beam sections are hidden from view by decking and inside flooring, except in the basement. The three-by-ten plate is bolted to the foundation by 8" × ½" lag bolts which are well cemented into the last course of concrete blocks. The author used three bolts for each ten-foot section of the plate, leaving the threaded end 3 inches above the top of the block foundation.

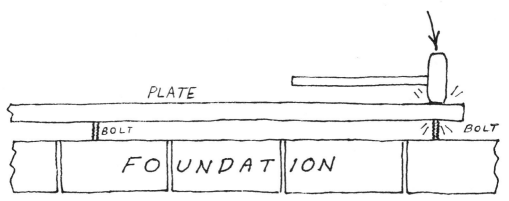

Fig. 26. Find out where to drill for the bolt holes by laying the plate upon the bolts and whacking the plank stiffly with a hammer. Then remove the plate and drill the half-inch holes where the bolts have left depressions in the wood.

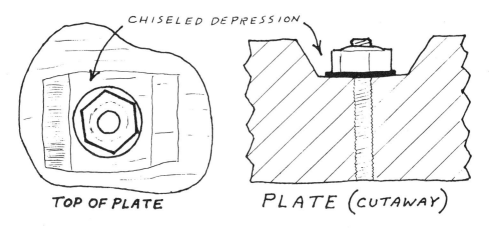

Fig. 27. It will be necessary to chisel enough wood out of the top of the plate to receive the washer, nut, and socket attachment.

place to hang pots, pans, mugs, and the like. Because the 8-by-10 ridge beam is firmly supported by four posts, and because the roof rafters are notched into the ridge beam as well as birdsmouthed upon the wall plate, the two tie-beams may not be strictly necessary. However, the wall plates are pieced together (two pieces on the east side, three on the west), and I am glad I installed the tie-beams for the extra rigidity.

Heavy framing with old barn timbers is just as physically demanding as building a traditional log cabin, so we measured and cut each section on the ground and then reassembled it in place with the help of friends — but there were only nine beams in our cottage that required help, compared with 60 or more long logs in a log cabin of comparable size.

Another drawback to the use of old timbers is that every one is of a slightly different dimension. In fact, the size will vary from one end of a beam to the other by as much as 1″. Obviously, a lot of extra measuring, figuring, and notching is necessary. This is tedious work, but worth it as it's good to be square and level when the time comes for putting up the roof rafters.

It may be desirable to build the roof before any cordwood masonry is laid up so that you'll be able to work in wet weather, and you give the cordwood extra drying time. The post-and-beam method is the only one of the three which allows this option, but if it is taken, temporary diagonals will need to be nailed up on the outside frame for rigidity until the cordwood is laid up. At our Cottage, the roof was completed before a single log-end was placed in the framework.

Ordinary framing, roofing, and roof-insulating techniques may be employed once the top plate is secure. It is beyond the scope of this book to enter into all phases of building. There are many good manuals on timber framing, plumbing, electric wiring, et cetera. I will not embark on these subjects, except where the cordwood masonry technique demands special considerations.

BUILDING WITH LOG-ENDS

As this subject is the heart of the book, I will go into some detail and not assume previous masonry experience on the part of the reader. Much of what will be said here is common to all three methods of employing cordwood masonry, but I will note where special considerations to the individual styles are important.

TOOLS. The required tools for mixing mortar are a good wheelbarrow, a hoe, a shovel, and a water bucket. For cleaning the barrow — a must as soon as it is empty — the same bucket and some sort of stiff brush will suffice (a worn whiskbroom works very well). It's a good idea to use a different bucket for carrying mortar. I use a 2′-square piece of plywood as a mortar board, but a wooden mortar board must be kept damp, especially on hot days, or it will drain the mortar of its moisture. For laying log-ends by the insulated mortar method described below, a small trowel, such as a pointing trowel, is useful. A regular brick mason's trowel with, say, an 8″ blade is also handy, especially where wider bedding is necessary. A good pointing tool can be made by slightly bending the last ½″ of a kitchen knife.

It is my belief that a wall built of split log-ends is inherently stronger than one built of rounds, but with the post-and-beam style we are not relying on the cordwood to be load-supporting so the rounds are best employed using this method, if rounds are specifically desired. I will use cylindrical log-ends in the illustrations for this part, leaving split ends for Chapter 4.

If cedar is used, a rasping tool is handy to smooth away the hairy edges of the end grain. These fibers get in the way of pointing and inhibit the formation of a good bond. We've found that a Dragon Skin rasping block is the perfect tool for the job. A circular sanding motion will save on Dragon Skin, but it is advisable to keep one or two extra sheets on hand (Figure 28).

CHECKING. Cylindrical log-ends, especially those over 4″ in diameter, should have fairly substantial checks in them from drying (see Fig. 29). I'm wary of log-ends that don't. The middle 3″ of these checks should be stuffed with fiberglass insulation or oakum. A screwdriver is useful for stuffing checks (see Fig. 30).

A chainsaw, hand saw, axe, splitting wedge, and hammer are all useful in eliminating pesky protrusions and splitting a log-end to fit into a particular space.

GLOVES. A very important "tool" for cordwood construction is a pair of rubber gloves. If gloves are not used, you'll wind up with cracked hands and fingers full of nasty "cement holes" (which take forever to heal) after 2 days of work. Jack Henstridge observes:

Figs. 28, 29, 30. (Left) Preparing the log-ends with a rasping block. (Top right) This check in a large cedar log-end has opened up more since laying-in. (Lower right) Stuffing the checks with fiberglass.

The most important piece of equipment you will need is a pair of rubber gloves, the kind used for housework, anything else is too bulky. You are going to use a pile of them because they tear quite easily. If you are right-handed you will end up with a pile of left-handed gloves, and vice-versa. No sweat. Take one of the good left-hand gloves and turn it inside out, now you have a right-hand glove.[6]

Rubber mason's gloves are very tough, but they take a lot of getting used to because of their bulk.

THE MIX. For use with cordwood as masonry infilling, I recommend either of the following equivalent mixes:

6 sand, 8 sawdust, 2 Portland cement, 2 lime

or,

6 sand, 8 sawdust, 3 masonry cement, 1 lime

The first mix, of course, is divisible by two, reducing it to 3 sand, 4 sawdust, 1 Portland, 1 lime. I provide the larger mix to compare it with the masonry cement mix and also because it makes a good wheelbarrow load, if rounded shovelfuls are used. All mixes in this book call for equal parts by *volume*.

I have experimented with quite a few different mixes, some of them without sawdust. My experience is that the non-sawdust mixes shrink and the sawdust mixes do not.

MIXING. Cordwood masonry is quite time-consuming with the post-and-beam method, as the builder is frequently trying to fit a piece against the framework. Smaller mortar joints can be used, as the strength of the mortar matrix is not as important as when the cordwood is load-supporting. This means that the mortar goes a long way. It is important that the mortar stay fresh. A powered mixer, then, is really no advantage with the post-and-beam style. Hand mixing is kinder to the environment, cheaper, and the mortar is easier to "fine-tune" for moisture content. The ingredients should be dry-mixed in a barrow (Figure 31), then the water is gradually mixed in. Soon, the consistency will be more conducive to mixing with a hoe. The bottom of the barrow should be scraped clean of dry sand and lime to assure a thorough mix (Figure 32). The initial mix is fairly wet and should be allowed to sit for five minutes to give the sawdust a chance to absorb moisture. This may dry out the mix to the extent that a little more water needs to be mixed in. The final mix should be damp, but thick enough so that the bedding will support the log-ends without spilling all over the ground. The mortar should hold its shape when placed halfway up on an adjacent log-end. If it doesn't, it's too thin.

INSULATION. Log-ends are good insulation but solid mortar is not, so we devised a method of insulating the mortar matrix. Our first job each day was to damp-brush the posts and beams to remove the dust. Then we tacked a 2"- to 3"-wide strip of fiberglass insulation — ½" thick is plenty — along the center of the bottom plate and up the posts (Figure 33). The strips were cut from batts or rolls of insulation with a utility knife. As we used 9" log-ends,

Figs. 31, 32. (Left) Dry-mix the mortar thoroughly with a shovel. (Right) Whip the mortar with a hoe. Scrape the barrow for dry material lurking in the corners.

this left room for a 3″ bead of mortar on each side of the insulation. There would not be room for the insulation strip on walls less than 8″ thick, but I wouldn't go less than that anyway, except perhaps in the Deep South, or for unheated outbuildings. In those cases, I would use a solid bed of sawdust mortar and leave out the fiberglass.

Fig. 33. First steps: Tack strips of insulation along the beams. Lay a double bed of mortar.

As the wall is laid up, the strip of fiberglass is continuously snaked over and under the log-ends, as shown in Figures 34, 35, and 36. Where three round log-ends meet, a wad of insulation is stuffed in to fill the gap. Tacking the strip along the posts keeps it from constantly falling into the work. We tried tacking a strip to the top beam, but it seemed to get in the way of the last course, making it difficult to get a tight fit without pushing out the insulation. We learned that it was easier to jam in the insulation at the top after all the log-ends were in place.

This method of insulating the mortar matrix was our own idea and has worked very well. It slows down the masonry work, to be sure, but a warm house makes the extra work worth it.

LAYING UP

It is very important, when laying up the first course, to get away from the flat plate and into a random pattern as quickly as possible. If the first course is laid with log-ends of the same diameter, there is a danger of getting stuck in a pattern which is hard to break and, paradoxically, hard to maintain. Aesthetically, masonry looks good if it is totally random or if it is very carefully laid up to a pattern. It looks bad if someone tried to incorporate a pattern and failed; and it looks bad if a seemingly random wall becomes patterned.

I like the random look: mistakes are less obvious, the wall is strong, and it incorporates all different diameters and shapes of log-ends. This latter feature helps ensure that the various sizes of log-ends in the pile are depleted at an equal rate, leaving a good selection right to the end. One of the little tricks of stone masonry, which applies equally to cordwood, is to keep a variety of pieces handy. Probability dictates that the right log-end is almost certain to be in a pile of, say, 50 random pieces, but sometimes, especially near the framework, it will be necessary to split a specific piece from a clear-grained log-end.

Good pieces to start with are log-ends split in half, or, for corners, split in quarters. Another handy shape is what I call a "slat-end." A slat is the first piece taken off a log when it is being squared for lumber. We acquired a couple of pick-up truck loads of slats at no cost when a friend was having cedar milled.

Figure 37 shows the importance of the first course. Once a random line is established, there will no longer be true courses. Rather, the mason will be dealing with individual spaces of varying sizes. The wall will almost build itself, calling out for the next piece required. This is as it should be for a truly random wall. Also, it helps to keep mortar joints small — ¼" to 1" is good. Any smaller makes for difficult pointing; any larger and it's hard to insulate, doesn't look good and wastes mortar.

When I suggest that the wall will build itself, this is not meant to imply that the builder can turn his brain off for the rest of the panel. Sometimes chance

52

Figs. 34 , 35 , 36. Fiberglass insulation is woven in and out between the log-ends, within a double bed of mortar.

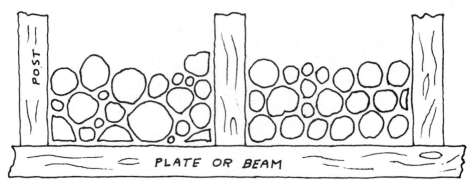

Fig. 37. The first course establishes the random pattern.

will throw the wall into an unwanted pattern. The first warning is when five or six log-ends of the same size appear in the same location. Log-ends of the same size fit beautifully into a hexagonal pattern, like a honeycomb, but this pattern will soon deplete one size very rapidly. If the builder happens to be using log-ends of consistent diameter, such as old utility poles, this may be desirable (see Cliff Shockey's home, pages 132–133).

Another exception to the "keep it random" rule would be if the builder deliberately wants to feature some planned configuration in the midst of a random background. This type of planned deviation can be extremely effective (see Fig. 132, page 142).

Another good reason for the mason to stay awake is that chance will rarely provide good opportunities to lay the 9″ to 12″ log-ends. Large ends require a "cradle," and this requirement will have to be provided for once in a while. This won't slow down construction, however, as one 10″ log-end gets a lot of wall up in a hurry (see Figs. 38 and 39).

Other points to remember during the laying-up:
• The builder should lean over the work once in a while, eyeballing it up and down, right and left. *Is the work plumb?* Large panels should be checked with a level now and again.
• The builder should stand back from the wall once in a while and ask himself, "Is the work balanced? Does it please the eye? Why? Why not?"
• *One mortar joint should not be placed directly over another.* This basic rule of masonry is even more important when building with log-ends because there is little bond between wood and mortar.
• The going gets tough near the top plate. It is a great help for someone to help the mason from the other side of the wall, by stuffing insulation, laying bedding, and guiding pieces into place. Sometimes, in close quarters, a log-end will only fit in from the other side. A good way to fill the final joint is to push the mortar off the back of the trowel and into place with the pointing knife.
• The builder should not be afraid to use odd-shaped pieces. They'll add interest to the wall.

54

Figs. 38, 39. Cradling . . . for a big one.

POINTING

This is one of the most important parts of the whole job. Many people think that pointing is purely decorative, to smoothen rough cement. Not so. Proper pointing greatly *strengthens* the wall by tightening the joints and reducing the chance of mortar cracks. The pressure of the mortar against the wood under the pointing knife gives the best chance for a good bond. Therefore, a fair amount of pressure should be applied to the knife as it is drawn along between log-ends, but not so much as to push it into the insulation cavity.

But, yes, pointing is *decorative* as well and is important to the finished appearance. There are many kinds of pointing and tools with which to point, but Jaki and I feel that the wood, and not the mortar, is the predominant feature of cordwood masonry, so we accent the wood with a simple recessed

55

pointing, the log-ends ¼″ to ¾″ proud of the mortar. Recessed pointing helps provide the interesting relief of a log-end wall. When using cylindrical log-ends, there is another mark against fancy pointing, such as raised V-joint: the joint is very small where the log-ends are tangent and wide in the space between three ends, so a time-consuming job becomes tedious and difficult.

Pointing is done while the mortar is still plastic. On hot days care should be taken that the pointing is not put off too long. Usually, there will be enough mortar squeezing out of the joints with which to point, but it is a good idea to save a little at the end of a day's laying in case there are substantial holes to fill. A really big hole should be stuffed with insulation first. Pointing is shown in Figures 40 and 41.

After work, tools should be washed and the work covered with a wet towel, especially the outside. This aids in seasoning the mortar slowly, thus preventing shrinkage. If the panel gets direct sunlight, the towel should be left on the next day, too. This is particularly advisable if a non-sawdust mortar is used.

Figs. 40, 41. Pointing a panel of cordwood masonry.

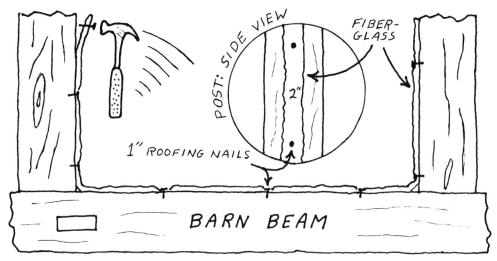

Fig. 42. Tack 2-inch-wide strips of fiberglass insulation that is ⅜ inch to ½ inch thick along the plate beam and up the sides of the posts to prevent its falling into your work.

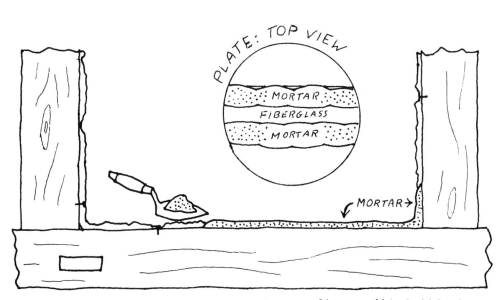

Fig. 43. Using a small trowel, lay a bead of mortar, ⅜ inch to ½ inch thick, along each side of the insulation on the plate beam and a few inches up the posts.

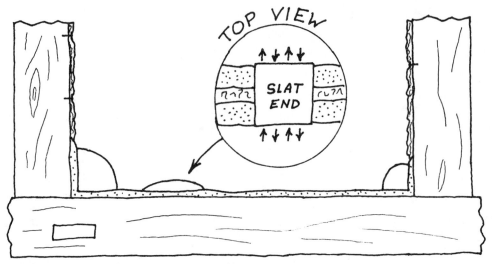

Fig. 44. Good pieces to start with are quarter log-ends and slat-ends. Lay them firmly in place, applying enough pressure so that the mortar squeezes out a little. A slight sliding motion back and forth at right angles to the plate will assure a good tight bond.

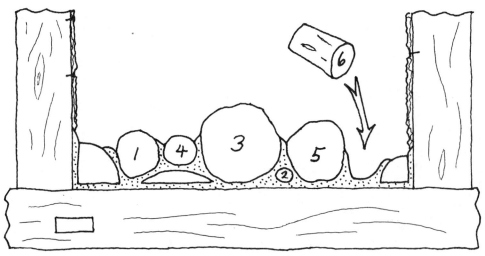

Fig. 45. Complete the first course, bedding and insulating up the sides of adjacent log-ends as necessary. Numbers indicate suggested order of placement.

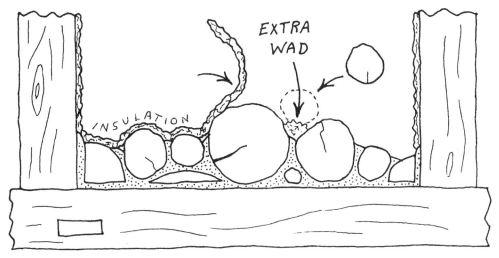

Fig. 46. To save time and labor, lay long strips of insulation whenever possible. Pack an extra wad in wherever a triangular space will be formed by a log-end capping two other log-ends.

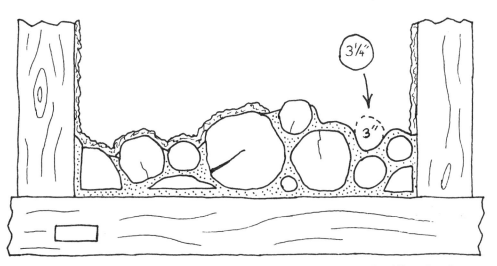

Fig. 47. Try to form the shape of the next log-end with your bedding, but just a tiny bit smaller than the piece you have in mind, so that it'll force a little mortar out, assuring the best possible bond.

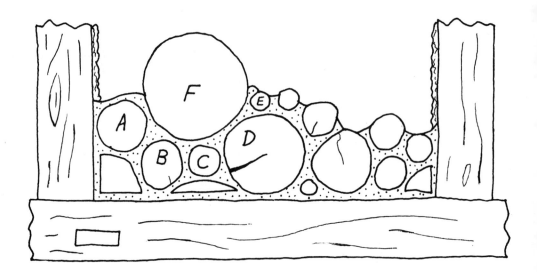

Figs. 48, 49. Plan "cradles" for your big log-ends, again, so that the cradle is just a wee bit smaller than the end you are going to lay up. Log-ends B, C, D, and E, with the mortar bed above them, form the cradle for the large end F. Use a little pressure and that slight sliding motion to squeeze the mortar out a little. Don't be too neat with your mortar. Make a bit of a mess. A mess of mortar on the ground is a sign of a good bond. If you set the mortar board under your work and up against the plate, you should be able to salvage most of the mortar that falls off the wall. You can catch the stuff on the other side with your hand or the trowel. Wear gloves. Don't pressure your log-ends so much that the one you are laying up squishes down and touches the one below. Leave at least a quarter inch of bed between ends where they're tangent; one half-inch is better. A protrusion on a log-end can often be fitted into the triangular space enclosed by three ends, but sometimes there is no way to avoid a protrusion touching an adjacent piece and establishing the gap between them.

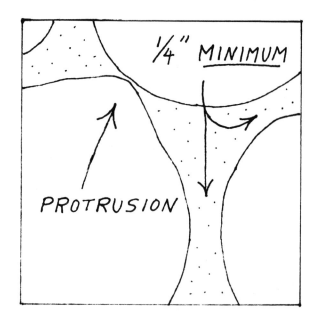

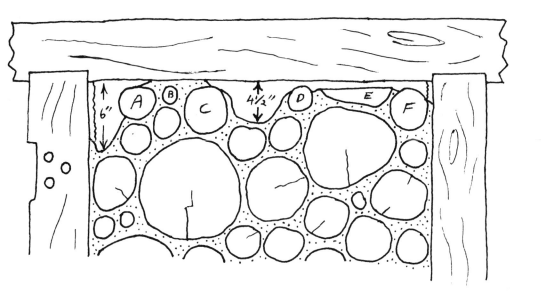

Figs. 50, 51. It gets tougher near the top. A, B, C, D, E, and F are easy to find if you have kept a good selection of small ends and slat-ends handy. The upper left-hand corner wants a slat-end, but to get a good fit, you split a couple of inches off one that is the right shape, but too big. The other space will be fitted nicely with half a medium log-end. After bedding, take a measurement before you split, allowing ½ inch for the topmost mortar joint.

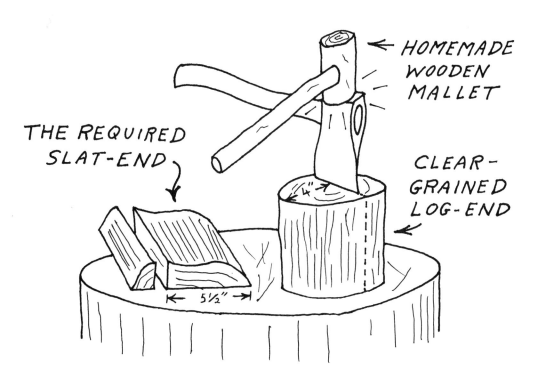

THE REQUIRED SLAT-END

HOMEMADE WOODEN MALLET

CLEAR-GRAINED LOG-END

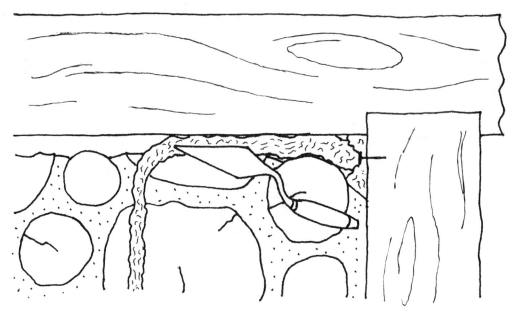

Fig. 52. Use the point of your trowel to stuff insulation over the last course.

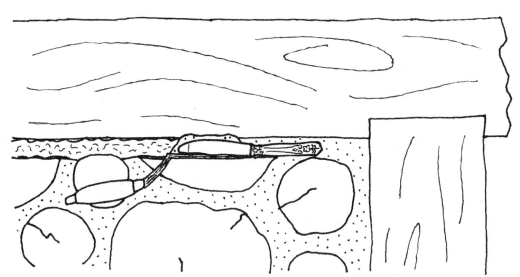

Fig. 53. To fill the final gap, push mortar off the back of the trowel with the pointing knife.

Fig. 54. The finished panel. The "beam-end" in the middle breaks the monotony of the round shapes. Pointing tidies up the panel and helps strengthen the mortar joints. When the house is finished, and the mortar is well seasoned, you can clean dirty log-ends and beams of any dusty mortar with a whisk broom. Particles of mortar do not adhere well to wood.

Fig. 55. A cordwood sauna.

Cordwood old and new: A wall of a 100-year-old Canadian cordwood barn (AI); and (A2) the home of Auguste Gallant and Bonny Pond, of Petit Rocher, New Brunswick, built in1978, combining the post-and-beam and the curved-wall styles (see Chapter 6).

Euclide Bourgeois built this story-and-a-half addition to his home (B1) in 1974 for about $1.50/sq ft, while Ferny Richard's pastoral hideaway (B2) near Rogersville, New Brunswick, cost him a little more than twice that amount in 1975. See Chapter 6.

B2

Cordwood old and new: A wall of a 100-year-old Canadian cordwood barn (A1); and (A2) the home of Auguste Gallant and Bonny Pond, of Petit Rocher, New Brunswick, built in 1978, combining the post-and-beam and the curved-wall styles (see Chapter 6).

A2

Euclide Bourgeois built this story-and-a-half addition to his home (B1) in 1974 for about $1.50/sq ft, while Ferny Richard's pastoral hideaway (B2) near Rogersville, New Brunswick, cost him a little more than twice that amount in 1975. See Chapter 6.

C1

Cordwood interiors: Mal Miller's 24″ walls created wide window wells with many possible uses (C1); C2 illustrates cordwood's rustic beauty—a wall of split red pine preserved with linseed oil.

C2

C

Two views of Jack Henstridge's famous "Ship with Wings,"
a 2600 sq ft home featuring an elevated geodesic-dome
living room and an airplane suspended from a gambrel
ceiling. This is truly a house for all seasons! See Chapter 6.

E1

Jean-Luc Chiasson's home (E1) is a combination of the post-and-beam style with log-ends used as load-supporting. Gottleib Selte's combination meditation *zendo* and temporary shelter (E2) is pentagonal in shape and was built in only three weeks. See Chapter 6.

E2

E

Tom and Helen Kwiatkowski's dome, built in 1979, illustrates what you can do with cordwood and imagination. A 5'-high kicker wall supports the geodesic-dome framework. The strength and beauty of one of the 12 built-up corners they employed is illustrated in F2. See Chapter 6.

F2

The author's "Log End Cave," an earth-sheltered house featuring 10″ cedar cordwood, built in 1976 (GI). G2 is a close-up of a wall of "Log End Cottage," the author's first cordwood structure, with the Sencenbaugh 500-watt windplant that provides electricity for the "Cave" in the background. See Chapter 6.

G2

H1. Cordwood structure with built-up corners, in Scottstown, New Brunswick.

H2. Bottle-ends in a wall-section of "Log End Cave." See Chapter 7, "Special Effects."

H3. Andy Boyd's huge Arkansas cordwood home, shown here in the process of being built.

H

BUILDING A POST-AND-BEAM SAUNA

To conclude this chapter, I offer complete plans for a very simple post-and-beam structure. It is designed as an outdoor sauna, but the plan could easily be adapted to a wilderness cabin or garden shed, or might even provide temporary shelter while a larger structure is built, to be used as a sauna or shed later on. Swedish owner/builders have for years adopted the strategy of building their sauna first (an important priority to the Scandinavian people), and moving into the building until their house is complete (Figure 55).

First, 8 cubic yards of sand is spread over a 12'8" by 17' area to a depth of 1'. The two volumes are equivalent. The *pad* thus created should be tamped while it is moist, to assure maximum compaction. The pad's purpose is to draw water away from the sauna, which is *floated* on the pad. This building technique, known as the *floating slab,* is one of the cheapest, easiest, and most effective means of building a frost-protected foundation for a cordwood structure. For a home, of course, the pad would be very much thicker, at least 18", as will be seen later.

FOOTING FORMS. A track with the dimensions shown in Figure 56 is dug in the pad with a flat shovel, and 2-by-8 forms are set in place as shown. Two pieces of No. 4 rebar (½" stock) are placed in the track, to be drawn up into

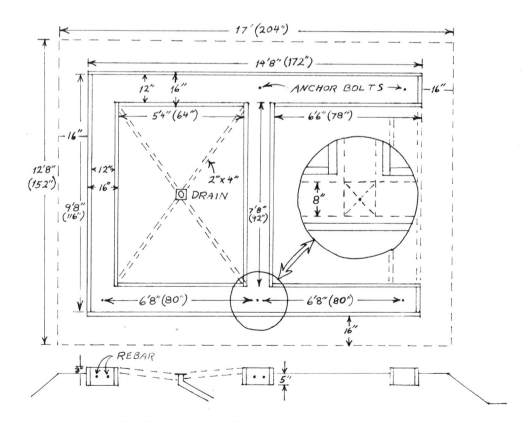

Fig. 56. Shuttering diagram for the footings.

the center of the footing as the concrete is poured. If care is taken with the forms, and they are cleaned after use, they can be used as plates or for any of several other purposes later in the construction. Anchor bolts should be set in each corner, where the posts will be. This eliminates the need for a floor plate.

As shown in Figure 56, a 4″ drain should be set into place in the pad before the forms are set in place. The forms should be oiled for easy removal and cleated with movable cleats as shown in Figure 57. This type of cleat guards against bulging between the corners. That they are movable makes for easy screeding of the concrete.

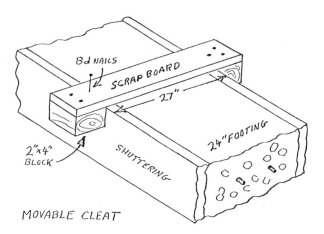

Fig. 57. Movable cleats will keep the shuttering intact during the pour.

CONCRETE. There are only 1.1 cubic yards of concrete required for the footings, so the concrete could be mixed in a wheelbarrow, although a mixer would make the job easier. The mix should be 1 part Portland cement, 2 of sand, and 3 parts crushed stone (¾″ diameter). Anchor bolts and rebar should not be neglected. After a day, the forms are removed, cleaned, and stacked aside.

THE FLOOR should slope from all corners to the 4″ drain in the center. This is not difficult to do. Four 2-by-4's, each measuring 4′6″, are set into the sand to a depth of 1″ (see Fig. 56). They should be set so that they start at footing level in the corners and slope downward about ½″ to the center. This slope can be checked with an ordinary level.

The floor, 3″ thick, can now be poured in four triangular sections. Each section can be screeded using the 2-by-4's as guides, so that water will be carried to the drain at the center. Wire mesh reinforcing is advisable, but not absolutely imperative. One-inch Styrofoam® under the concrete floor is another option, one which will assure a warmer floor. If Styrofoam® is not used, it would be a good idea to use sawdust concrete for the same purpose. It is made by replacing the crushed stone in the previous mix with an equal volume of sawdust, the kind that comes from a sawmill. The ingredients should be dry-mixed and then the water should be added. As with sawdust mortar, the mix should sit for five minutes to allow the sawdust a chance to soak up moisture, then a little more water should be added and the stuff remixed to a

66

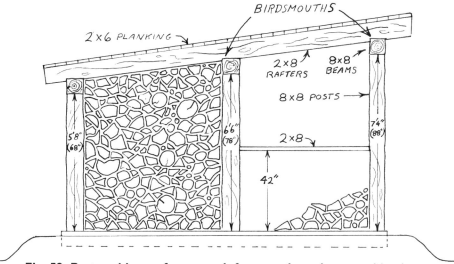

Fig. 58. Post-and-beam framework for a cordwood sauna, side view.

workable consistency. Because of the sand pad, a plastic sheet under the floor is not really necessary. There will not be a water problem.

FRAMEWORK. Next, the framework can be built. Six 8-by-8 barn timber posts are stood up at the corners, with heights as shown in Figure 58. Holes drilled in the bottom of the posts receive the anchor bolts. Three similar 10′ barn beams span the three sets of posts, allowing a 4″ overhang by the beams as an aesthetic consideration. Temporary diagonal braces should be nailed to the framework for rigidity until the roof and cordwood masonry are complete. Six 16′ 2-by-8's make up the rafter system (Figure 59). They are birdsmouthed onto the support beams as shown in Figure 58.

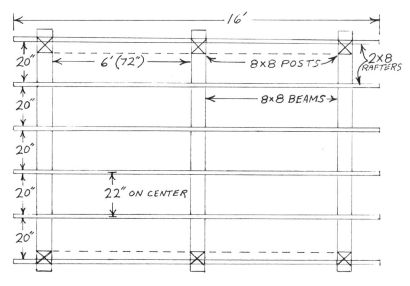

Fig. 59. Roof support system for the cordwood sauna.

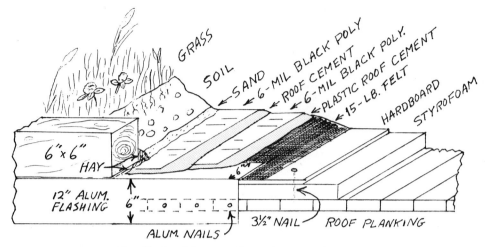

Fig. 60. Roof layers of a sod-covered sauna.

ROOF. Planking is made up of 12′ rough-cut 2-by-6 boards, planed on one side. After the planks are nailed up, their ends may be trimmed to a straight edge with a chainsaw or circular saw. The various roofing layers can now be applied (Figure 60).

I recommend that the whole roof be covered with 2 inches of Styrofoam®, in case you'll want eventually to close off the sun porch for wintertime use. Six sheets of 4′ by 8′ Styrofoam® will do the trick. A layer of hardboard is placed on top of the Styrofoam® to protect it during construction of the earthen roof and to supply a good surface for the application of the water-proof membrane. The cheapest available hardboard should be used. Chip-board, recycled plywood, or even Masonite would all do the job. The hard-board can be fastened to the roof with 3½″ nails. Fifteen-pound felt paper is stapled to the hardboard, beginning at the lower part of the roof and laying each succeeding course to lap the previous one by 4″. Six courses of paper will complete this layer, the purpose of which is to allow movement between the membrane and the roof.

Flashing is applied next. The edge of the roof can be flashed with 50′ of 10″ or 12″ aluminum flashing. The flashing should be folded down the middle, so that half hangs over the edge and half rests on the roof. The flashing can be stapled to the hardboard, as there will be a layer of black plastic roofing cement directly over it, but the sides should be nailed to the ends of the plank-ing with aluminum nails having rubber washers beneath the heads. With ordinary roofing nails there is the danger of electrolysis eventually eroding the aluminum. If galvanized nails must be used, put a bead of silicon caulking over each nail. One nail per plank is enough. The 1″ or 2″ overhang (depend-ing on the flashing used) will act as a drip edge and keep the water away from the underside of the planking.

This is a good time to cut the hole in the roof for the stovepipe. Location measurements should be carefully transposed from the interior to the roof. The stove and stovepipe must be kept a safe distance from combustible ma-

Fig. 61. A view of the roof layers prior to the application of the waterproof membrane.

terials. I believe in Metalbestos insulated stovepipe. It is expensive, but will last for the life of the building. Metalbestos is safe within 2″ of combustible material, so a hole should be cut with a diameter 4″ greater than that of the stovepipe. The best tool for the job is a chainsaw, but take care to hold the saw tightly while cutting through the hardboard with the tip. Figure 61 shows the planking, Styrofoam®, hardboard, and felt paper.

WATERPROOFING THE ROOF. For a small roof like this one, I recommend the 6-mil black polyethylene waterproofing that we used at Log End Cave. As I said in *Underground Houses,* I would not want to do a large roof by this method again as it is a thankless dirty job, and it's hard to wrestle with a sheet of plastic 32′ wide and 100′ long. For the sauna roof, however, two 12′ by 16′ sheets of polyethylene will do. One day's dirty work will waterproof the roof.

The technique is to start at the top of the roof and apply a 2′-wide coating of black plastic roofing cement — also called salvage cement or double-coverage cement — along the top. Place the 6-mil black polyethylene on the roofing cement so that it can be unfolded down the roof as work progresses. It would be a good idea to unfold and refold the sheet after it is bedded into the first area of cement, to check that it is being applied squarely. The process is repeated down the roof to the lower edge. The first layer of plastic should lap the flashing by about 2″. The second and final layer should lap the first by 2″, thus *feathering* the edge. An inch or two of the flashing will be exposed on the edge of the roof. The second layer is the insurance layer, and is also bedded in cement (see Fig. 62).

A retaining wall for the earth can easily be made from barked cedar logs or treated landscaping timbers (4″-by-4″, 6″-by-6″, etc.). They should be well spiked to each other at the corners and should be placed on a 1″ bed of hay, which will protect the membrane and finally decompose into a fibrous

Fig. 62. Applying the waterproof membrane.

filter for keeping the rain runoff clean. A little extra hay can be tucked next to the logs for the same reason, prior to the application of the earth.

THE SOD ROOF. The roof should be allowed to grow wild. Unlike an earth-sheltered house, it would be very difficult to haul a lawnmower onto and off this roof. Also, the long grass will act as a natural mulch, protecting the roof during a drought. I recommend that a 1″ layer of sand be applied to the whole roof before the soil is brought up. The sand will protect the membrane during the sodding, and facilitate roof drainage.

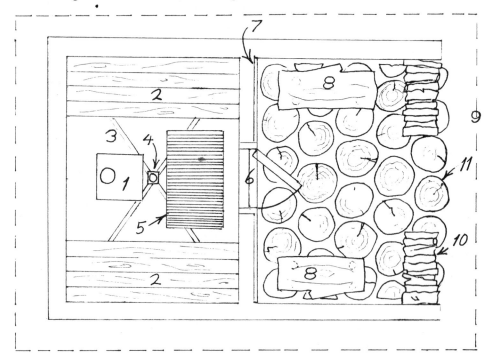

Fig. 63. Floor plan of cordwood sauna. (1) Stove; (2) Benches; (3) Two-by-fours set into the floor for screeding the slope towards the drain; (4) Drain; (5) Hardwood slats nailed onto a frame. This provides a place to stand and wash; (6) Door; (7) Thermopane windows; (8) Antechamber benches; (9) Extent of roof overhang; (10) Firewood; (11) Slabwood floor.

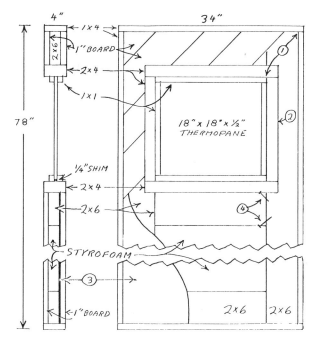

Fig. 64. A good door for a sauna.

① SILICONE CAULKING AT ALL THRU JOINTS
② 2"x6" CHECKED OUT FOR WINDOW FRAMING
③ 15-LB FELT UNDER OUTSIDE SHEATHING
④ 2"x6" FRAMEWORK HELD WITH TOENAILS UNTIL
     1" SHEATHING IS FIXED

The roof structure described in this section will support a 6″ sod roof with a 4′ snowload. The two choices are to plant the roof in place, or to haul cut sods up and kick them into place with the side of the boot. The easiest way to plant the roof is to build the soil up to the required depth and spread the chaff from the hay over the top. Then spread a ½″ layer of hay over the chaff and water the whole thing down. The hay will act as a mulch through which the new grass can easily grow. If sods are to be brought up, enough topsoil should be spread so that the addition of the sod will bring the depth up to 6″.

WINDOWS should be thermopane (double-glazed) and framed by 2-by-8's. More on windows later in the book.

THE DOOR should be homemade and well-insulated. A store-bought, finished door would really detract from the rustic appearance of this structure. People who make their own doors like to do it their own way, but I do include a cutaway view of the door which we used at Log End Cave as an example (Figure 64). This would make an excellent sauna door.

Let's see now, what have we forgotten? Of course, the walls! They are of cordwood, of course, as described on pages 57 through 63. The only special consideration applicable to a sauna would be to have one log-end, 4″ in diameter, near the base of the back wall which can be removed to supply air to the woodstove. Similarly, a couple of larger ends could be removable at eye level, for ventilation or to peek out the side walls. Removable log-ends should have a slight taper, be coated with vegetable oil before laying up, and carefully worked loose before the mortar is completely hard. Handles could be screwed on, if required.

Benches should be made of non-resinous wood of low conductivity. Poplar is perfect. Hardwoods tend to burn the baby's bottom. The benches can be made of four 2-by-6's cleated together from below. Nails should not be exposed. A good idea is to make the benches adjustable so that they can be set at either the 24″ or the 42″ level, or can even be set to a gentle slope for reclining.

The sauna is heated with a woodstove. Yes, an electric heater made especially for saunas could be used, but a lot of the romance would be lost. Care should be taken that safe clearance is maintained from combustible walls (*both* stove and stovepipe). If this puts the stove too far out into the clear space, insulated wallboard made for the purpose can be used behind the stove. There should be a 1″ airspace between the wallboard and the cordwood wall.

The antechamber, which should face south, is used as a dressing room and relaxation area between sessions in the sauna. The floor for this area could be 3″-thick slabs of wood — very short log-ends, really — set into the sand. If both sides are sealed with two or three coats of polyurethane, they will last a long, long time. Short cordwood walls act as a windbreak and help this area serve as a sun trap. If desired, the antechamber could be screened in for summertime use and eventually closed in for year-round use. If this is anticipated, it would certainly be worth going through the little extra work required to complete the footing between the two posts at the south end of the structure. Removable Plexiglas windows could be used to cover over the screened areas. A sliding glass door could be incorporated between the two southernmost posts. But we're getting kind of fancy. If kept simple, the sauna can be built quickly and for very little money.

4. Built-Up Corners

With the built-up corners method, the cordwood masonry itself does all the load supporting. Corners are built first so that a mason's line can be stretched between them to assist in keeping the walls plumb.

There are several ways to build up the corners. The Northern Housing Committee (N.H.C.) of the University of Manitoba suggests laying wood blocks with mortar, as with the rest of the wall. They build cordwood walls 2' thick to deal with the heating requirements of homes in the Canadian West. Their corner blocks, then, are 30"-long 8-by-8's, milled on three sides. Figure 65, reproduced with permission from N.H.C.'s book *Stackwall: How to Build It,* shows construction detail based on this method. With three sides finished, the corner will be strong and the outside edge is handy for attaching a corner clip for holding the mason's line. Malcolm Miller, of Fredericton, New Brunswick,

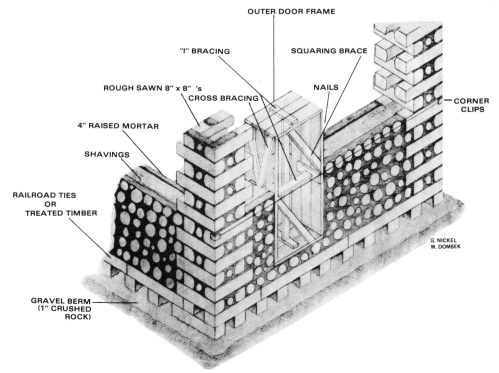

Fig. 65. Construction detail based on the built-up corners method.

Fig. 66. Stackwall barn in Hemmingford, Quebec, Canada, built in 1952.

built his corners with full-sized 6-by-6 timbers, pressure-treated for extra long life (see page 137). Rough-cut 6-by-6's are quite commonly found in building supply yards. Given a choice, the builder should try to obtain the driest ones.

Even though the bond between wood and mortar is not very great, these mortared-up corners are incredibly strong. I recently visited a stackwall barn in Hemmingford, Quebec, built in 1952 (shown in Figures 66 and 67). The structure was built with 10″ cedar log-ends that had not been barked. The corners were built with the same type and length of log-end as the wall, 3″ to

Fig. 67. Corner detail of the Hemmingford barn.

4″ in diameter. Like the log-ends, the corner pieces were cylinders for the most part. The structure was built without any special care taken to keep log-ends from touching one another and there is a dead air space in the mortar matrix by way of a thermal break. All in all, I would not have expected this building to have lasted as long as it has — 28 years — but, after inspecting the barn I am convinced that the walls will still be sound 28 years from now . . . and then some. In other words, a corner built with short, round, unbarked cedar log-ends, some of which are actually touching each other, will last at least a half-century. If care is taken, such as exercised by Malcolm Miller and the N.H.C. in their structures, the buildings will last hundreds of years, assuming a good foundation and periodic roof maintenance.

In an article appearing in *Farmstead #18,* I offered three alternatives to folks who are unable to obtain seasoned, squared timbers for the corners: 1) Use "beam-ends" cut from old barn beams in the built up corners. (Note: The use of creosote on externally exposed barn beams will give a pleasing contrast to the cordwood masonry and preserve those old timbers for a lot longer. Creosote should *never* be used indoors, as the obnoxious smell will persist for years.) 2) Spike green timbers together with 10″ spikes, the way many people are building "traditional" log cabins these days. Figure 2 shows a barn in Wisconsin that looks like it may have spiked rather than mortared corners. 3) Scrounge some dry recycled 2-by-6's and lay them up as shown in Figure 68. This drawing assumes a wall thickness of 16″. Tom Kwiatkowski, of Point au Roche, New York, adapted this idea very successfully to a twelve-sided kicker wall for his geodesic dome (see pages 118–119, and F).

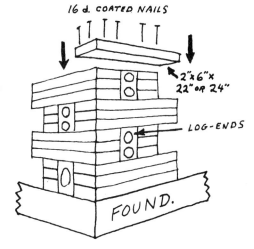

Fig. 68. A built-up corner made from old two-by-sixes.

I would like to add that seasoned log-ends, split into quarters, would also do very nicely as blocks for building up corners. The two flat sides should be kept down and out. Again, they should be 6″ or 8″ longer than the other log-ends, so that they will be tied into the wall.

No matter what material is used in the built-up corners, the technique for laying up the walls is the same. The corners are built up about 2′ or 3′, then a mason's line is stretched between two corners and the cordwood masonry walls are infilled by methods similar to those described in Chapter 3. Obviously, it is very important that the corners are built plumb in both directions. A 4' level is the best tool for this purpose, although a 2' level will do in a pinch.

Although the barn in Hemmingford and many others like it were successfully constructed with walls 10″ thick, Jack Henstridge and I agree that 12″ should be considered as the minimum thickness for a load-supporting wall in a dwelling. Sixteen-inch walls are preferred, especially in cold climates. A 24″ wall, such as Malcolm Miller's (page C), is 50 percent better in terms of insulation, but is not strictly necessary for load-supporting in a single-story or story-and-a-half structure.

Outside of wall thickness, there are two other areas where the building technique differs in a cordwood wall that is load-supporting: mortar mix and mortar thickness. I recommend decreasing the sawdust content slightly to produce a stronger mix:

6 sand, 6 sawdust, 2 Portland, 2 lime

or

6 sand, 6 sawdust, 3 masonry, 1 lime

These mixes are equivalent because masonry cement is pre-mixed with an approximate 2:1 ratio of Portland cement to lime.

In addition to a stronger mix, it is stronger to go with wider mortar joints than those used for masonry infilling. Joints should be between ¾″ and 1″ (⅝″ should be considered as the minimum). While discussing strength characteristics of a cordwood wall, I would like to reiterate that a wall of split ends will be inherently stronger than one of rounds. But rounds will do, and very nicely, if attention is paid to the basic principles of cordwood masonry described in this book. An incidental advantage of the larger mortar joints is that fewer small-diameter ends will be required than with smaller joints, speeding up the work somewhat.

The extra thickness of walls constructed by the built-up corners method eliminates concrete blocks as a possibility for a foundation, unless they are laid with the ends showing, as suggested for the temporary structure in Chapter 2, or in the particular case of building a 12″-thick wall on a foundation of 12″-wide blocks. Note that in Figure 65 the foundation is composed of railroad ties supported by a gravel berm. The berm recommended by the N.H.C. for a 2'-wide cordwood wall is shown in Figure 69. This is a good inexpensive foundation for anyone with access to old railroad ties.

A foundation that I am particularly enthusiastic about is the *floating slab*, very much like that described previously in the discussion of the sauna. The sand or gravel used must have excellent percolation qualities (it should not hold water) and the pad should be compacted in layers and built to a minimum depth of 18″. The pad's dimensions should be at least 6' greater than the external wall dimensions. As a protection against erosion, grass should be encouraged to grow on the 3'-wide berm thus created around the outside of the house. A thick layer of woodchips is another means of arresting erosion of the pad.

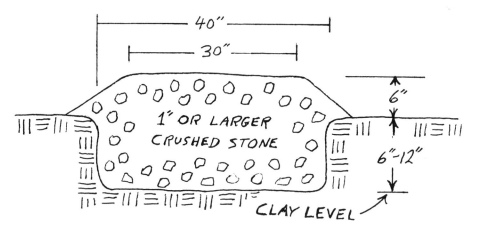

Fig. 69. Gravel berm foundation.

Note: It is important that organic material, such as sod or loamy topsoil, be scraped away from the site before you bring in the sand or gravel for the pad.

The footings and floor can be poured at the same time, if external dimensions do not exceed 40'. If either dimension is greater than that, the footing should be poured first and the floor should be poured two days later with a flexible and waterproof expansion joint, such as ¼" neoprene, between the floor and the footing (see Fig. 70). Floors should be 4" thick, reinforced with mesh, and have 1" of Styrofoam® beneath as insulation. The Styrofoam® can be protected with 1" or 2" of sand before the pour is made. Footing dimensions are a function of wall width. Assuming an 8'-high wall in an area of high snow load, the following footing dimensions are recommended.

Wall Thickness	Footing Dimensions		Number of Rebars
	H	W	
12"	10"	20"	2
16"	12"	24"	2
24"	12"	32"	3

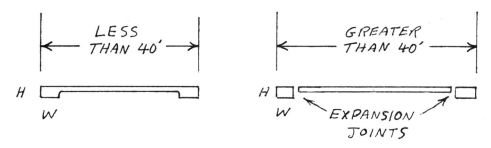

Fig. 70. If external dimensions exceed 40 feet, the footing should be poured first and the floor should be poured two days later.

There is no gain in making footings deeper than 12″. Additional concrete is better used in making the footing wider, thus spreading the weight load over a greater area. Half-inch rebars should be wired into the footing, supported about halfway between the top and bottom edges of the forms. Keep 8″ to 10″ between rebars.

The floating slab thus created offers several advantages over other foundations in areas with deep frost conditions. First, the pad of percolating material is the best protection against frost heaving. Second, much less concrete is used than on foundations requiring excavations to a 4′ depth or more. Third, it is fast and easy, and well within the capabilities of the owner/builder. Finally, the insulated floor acts as a thermal mass, helping to maintain a more constant temperature during the winter.

I know that some people have an aversion to concrete floors — I'm quite fond of wooden floors myself — but I think that a good part of this aversion stems from a presumption that concrete floors are cold and damp, probably because basement floors often display these characteristics. If insulated and built on a percolating pad as described, a concrete floor is neither cold nor damp. Jaki and I do not experience the slightest discomfort from the concrete floor in Log End Cave, although I would certainly go with 1″ of Styrofoam® next time so that the floor would operate at a higher temperature as a thermal mass.

WINDOWS AND DOORS

One of the unexpected beauties of walls that are 16″ to 24″ thick is the wide window wells, ideal for sitting, growing indoor plants, or starting garden seedlings. Windows can be framed as shown in Figure 65. The walls in this diagram are assumed to be 24″ thick, so the frame is constructed of three 2-by-8's (four 2-by-6's would also do). The cross-bracing is scrap 1-by boarding and is left in the wall. Other temporary bracing should be 2-by stock. When the walls are hard, the temporary bracing is removed and the window frame will hold its shape. Door frames can be constructed in much the same way, or the builder might change to the post-and-beam style just to frame the doors, as shown in Figure 71. I believe that the use of heavy posts for door framing would be a safeguard against potential frame twisting in the other method.

The N.H.C. recommends lintels over doors and windows, and sills beneath windows. The lintels can be a pair of 6-by-6 timbers installed just above the window or door frame. One of the lintels should be kept flush with the interior surface of the wall, the other flush with the exterior of the wall. Thus, they are exposed and are a pleasing design feature on both masonry surfaces. The lintels should extend 10″ to 12″ into the masonry on each side of the door or window. Their purpose is to pass the weight load of the roof onto the ma-

Fig. 71. Door framed with heavy posts.

sonry walls, without placing undue pressure on the window or door frames. The sills that the N.H.C. recommends beneath the windows are built in much the same way as the lintels (see Fig. 72).

Jack Henstridge used neither lintels nor sills with the windows of his house. My feeling is that the lintels are very important structurally, and the sills somewhat less important. Whether or not sills are used, the outermost bottom board of the frame should extend 2″ proud of the wall, and a drip edge should be cut into it with a chainsaw. Also, this board should have a very slight slope away from the window, so that water does not stand on the window frame (see Fig. 73).

Fig. 72. Lintels and sills.

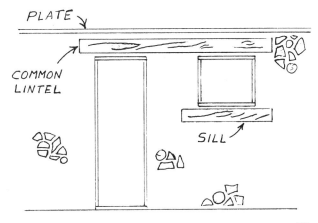

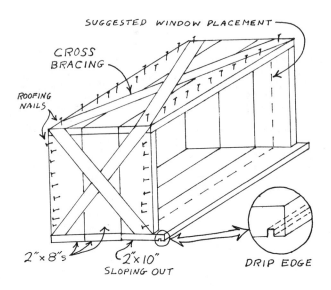

Fig. 73. Box frame for windows, showing the drip edge.

INSULATION

Previously, we discussed the use of fiberglass strips woven in and out of the mortar cavities as insulation. There are other options, and the extra wide, load-supporting walls make these options more feasible. The N.H.C. recommends sawdust, with one shovelful of dry lime added and mixed per wheelbarrow of sawdust. The lime protects against insect infestation and fungal growth.[7] The sawdust should be poured in and packed after every course. The N.H.C. recommends 3″ mortar beds to maximize the insulation value of the sawdust-filled cavity. I am inclined to disagree slightly in favor of mortar beds that are 4″ wide for greater wall strength. The sawdust that I use in my mortar mixes should offset the loss of 2″ of sawdust insulation.

Other loose-fill insulations, such as vermiculite or shredded beadboard, could be used in place of the sawdust-and-lime mixture. The use of sawdust, however, is probably the most economic insulation to use, both in terms of dollar-cost and energy-cost.

PLATES

As the weight of the wall puts an outward stress on the corners, it is important that the corners are tied together with plates. The plates are also necessary for fastening rafter or floor-joist systems. They provide a flat surface upon which to work and distribute the momentary load over the cordwood wall. Without the plate, rafters with a heavy load will be trying to "slice" into the wall. If it's not possible to span from corner to corner with a single plate, it is advisable to nail a double plate together in two courses. Plate material can be rough-cut or planed 2-by-4's or 2-by-6's and they should be spiked into the built-up corners. The N.H.C. recommends that 5″ spikes be driven into the plate every 2′, and the ends bent on the underside to act like anchor bolts. They advise that the plates be set into 2″ or 3″ of mortar and then tapped level.[8] Figure 74 shows the corner detail of a strong plate system.

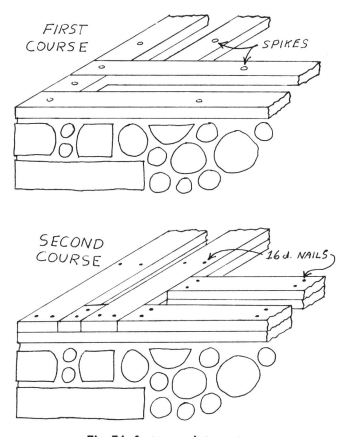

Fig. 74. A strong plate system.

Note that there is an inner plate and an outer plate. If 2-by-12's are used on a 24″ wall, or 2-by-10's on a 16″ wall, the inner plate may be omitted, though the interior appearance will not be as "finished." On a 12″ wall, 2-by-12 plates would be excellent; they'd be strong and would finish off the top of the wall beautifully. The owner/builder should make the best use of the available materials in deciding what plate system to employ. Cost and the degree of interior finish desired are two important determining factors to consider. I tend to "overbuild" slightly when making this kind of decision. Low construction cost and structural integrity are not mutually exclusive in an owner-built house.

Some house plans may call for two walls to join part-way along one of their lengths. The technique is demonstrated in Figure 75. Jack Henstridge says, "On a joining wall such as this, it isn't necessary to put in more than, say, three or four sections of tie blocks in an 8′ wall."

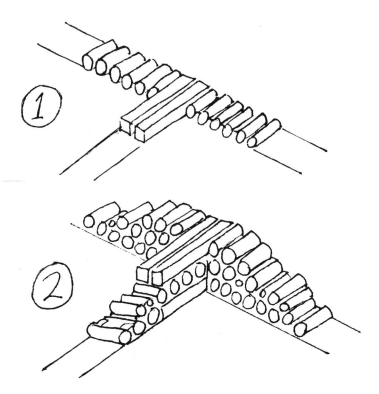

Fig. 75. Joining two cordwood walls by the use of tie blocks.

In other designs, it may be necessary to terminate a wall without a corner for some reason, as on a carport. Thanks to Cordwood Jack for the following advice and illustrations (Figures 76–84):

Step One. Construct a reference post at the end of your wall. Make sure it is securely in place, as it must stay there until you are finished. This isn't entirely necessary, but it will definitely simplify the job.

Step Two. Lay the first course of blocks in the normal manner, but keep in mind that you are going to be laying the tying blocks at 90° to them at the end of the wall, so try to keep those end blocks as close to "size" as possible.

Step Three. Put in the tying blocks. A slight downslope into the wall is okay. Try not to slope them the other way, though; the reason is obvious.

Step Four. Just keep repeating the procedure right to the top of the wall, then remove the reference post. The completed wall is shown in Figure 80.

Well, that's about it for built-up corners. They are beautiful and strong, and they are the only way to use cordwood as load-supporting in a building with corners. And they don't have to be square corners, either, as Tom Kwiatkowski's twelve-sided home illustrates (pages F, 119).

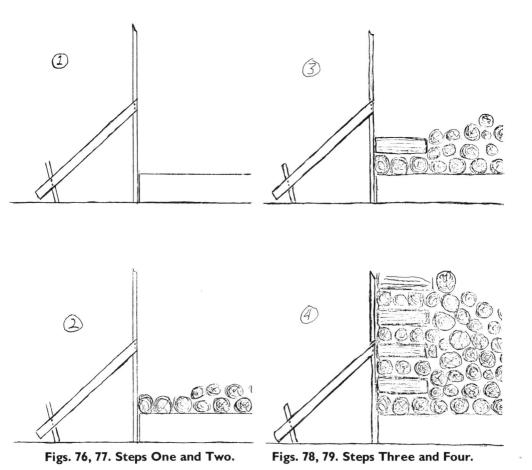

Figs. 76, 77. Steps One and Two.   **Figs. 78, 79. Steps Three and Four.**

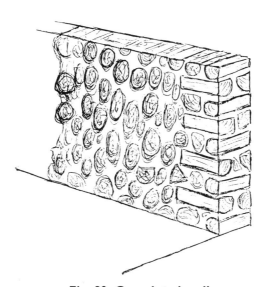

Fig. 80. Completed wall.

Fig. 81.

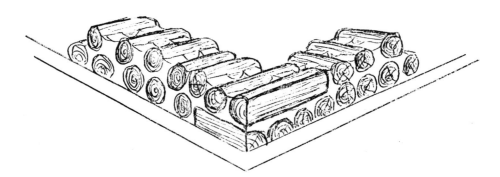

Fig. 82.

Fig. 83.

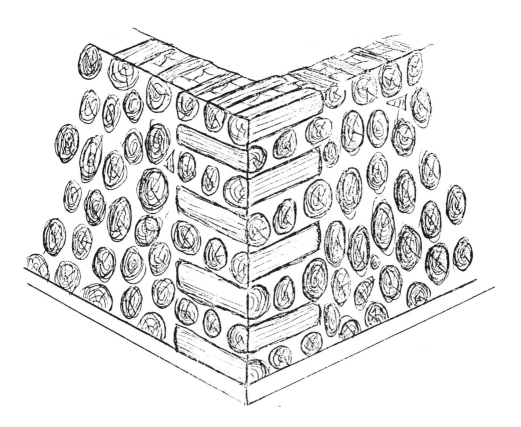

Fig. 84.

5. Curved Walls

"But who needs square corners?" I can hear Jack Henstridge screaming all the way from New Brunswick. "Avoid them like the plague; they are evil. The Boogey Man can catch you there. Besides that, they are more difficult to build. Also, by curving the wall you develop lateral strength. To illustrate what I mean, take a piece of writing paper and glue the top to the bottom, making a cylinder. Stand it on one open end and put a plate on top. Now start piling stuff on the plate. You will be amazed at how much it will hold. The same applies to a wall: curve it and you strengthen it."[9]

As far as laying up the individual log-ends is concerned, there is very little difference between the curved-wall method and the built-up corners method: the mortar mix is the same and the wall thickness recommendations are the same. About the only difference is that the mortar joints will be wider on the outside of the curved wall than on the inside. Of course, this difference is greater on thicker walls than on thinner walls. Similarly, the difference is greater on small diameter houses than on larger. Here are six examples which illustrate these relationships:

House	Wall Thickness	Outside Diameter	Inside Diameter	Difference in Mortar Joints
A	12″	32′	30′	⅜″
B	16″	32′8″	30′	½″
C	24″	34′	30′	¾″–1″
D	12″	40′	38′	¼″
E	16″	40′8″	38′	⅜″
F	24″	42′	38′	⅝″

It is important that the inner joint be thick enough to aid in load-supporting, so the wall should be built with the log-end spacing established on the *inside*. The outer joints will be thicker as shown in the chart. This chart assumes a log-end with an average diameter of 5″. If smaller ends are used, the difference in size between inner and outer joints will be smaller, and if bigger ends are used, there will be a greater difference.

The difference between mortar joints only occurs in a lateral direction; top-to-bottom jointing stays the same. The effect is shown, slightly exaggerated, in Figure 85. Because of the resulting wide outside joints — about 2″ on the

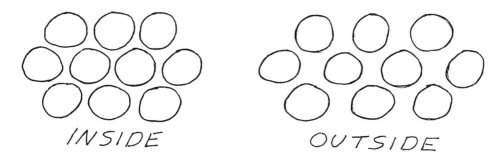

Fig. 85. The effect of a thick wall on mortar joints.

average — I would be disinclined to use 24″ log-ends on a 30′-diameter building. Such wide walls would be unnecessary on a small structure, anyway, as the home would be very strong and easy to heat with 16″ ends.

Laying up a curved wall of cordwood is actually much easier than laying up the walls of a rectilinear structure. For a start, there are no posts or built-up corners to fit up to; only door and window frames will slow down the masonry work. What's more, the walls will remain absolutely round, plumb, and flush on the interior by using the method described below.

THE ROUND HOUSE

To illustrate the curved-wall technique, we will walk through the planned construction of a single-story round house (see Fig. 86), that I have designed. The house has an outside diameter (O.D.) of 38′, and 16″ cordwood walls, giving an inside diameter (I.D.) of 35′8″. The gross area, then, is 1134 sq ft, while the interior area is 999 sq ft (let's call it a thousand). Notice that 135 sq ft are lost to the thick cordwood walls. This is why it's important to deal in terms of *usable interior square footage*. This should be pointed out to the tax assessor. Figures 87, 88, and 89 show, respectively, the rafter plan, the floor plan, and a cross-sectional view along the north-south axis.

THE FLOATING SLAB

The house is designed to be built on a floating slab. The building site should have fairly good drainage to begin with; if it does not, a considerable amount of fill will have to be brought in, or the house should be built on piers. If drainage conditions are only fair, I would build the pad 24″ thick instead of 18″. The site should be scraped with a bulldozer to remove all organic

Fig. 86. The Round House.

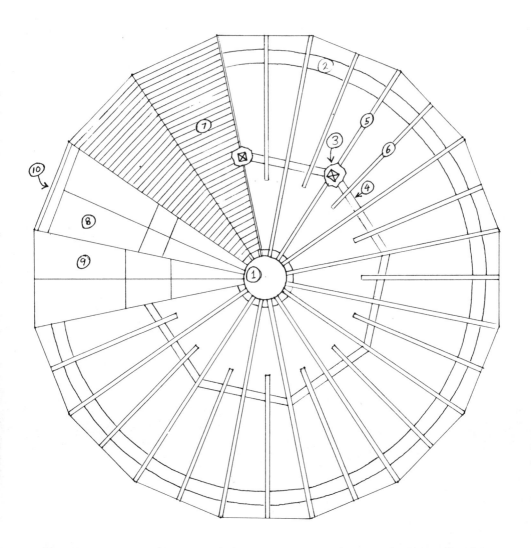

Fig. 87. Structural plan of the Round House. (1) Stone heat sink; (2) 16-inch cordwood wall; (3) Post locations, eight in all; (4) Eight-by-eight beam, part of octagonal support structure; (5) Primary rafter; (6) Secondary rafter; (7) Two-by-six planking; (8) Rigid foam insulation; (9) Hardboard; (10) Two-by-four or four-by-four, to match depth of insulation.

material such as sod, leaf mold, or rich topsoil. This is to safeguard against a settling of the structure. The pad material should have excellent drainage characteristics; good gravel or clean sand are ideal. It is easier to drive forming stakes into sand, but gravel offers better support under heavy loads.

The diameter of the pad should be about 8′ greater than that of the footings. In the 38′-diameter house, the approximate diameter of the pad is 46′. Using the πr^2 formula, we find that the area of the pad will be: 3.1416 × 23 × 23 = 1,662 sq ft. Dividing by nine yields square yards: 1,662/9 = 184.67 sq yd. As the pad is 18″ (half a yard) thick, dividing by two will yield 92.33 cu yd of material needed to build the pad. (A 24″ pad will require 123 cu yd of sand or gravel.)

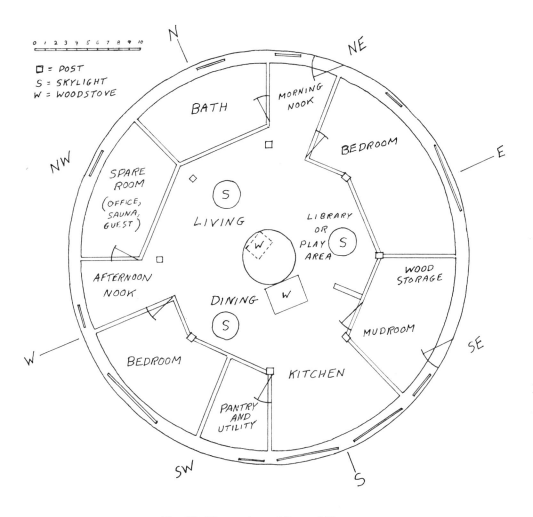

Fig. 88. Floor plan of Round House.

If the site already has good drainage and the subsoil yields good results on a percolation test, the pad can be built quickly with a bulldozer. No need to bring in any material in this happy circumstance.

The pad location and depth can easily be marked out for the 'dozer driver by choosing the center point of the building, driving a stake in the ground, and, with a 22' rope and a pointed stick, describing a circle around the site. If fourteen stakes are driven into the ground about every 10' around the 138' perimeter, the operator will know where to spread the sand or gravel. If the stakes are driven so that exactly 18" shows above mean grade, they will also serve as depth gauges. The center stake should also be marked or cut to establish the new grade 18" above the grade established by the scraping of organic material. These reference stakes should be checked with a surveyor's level, a tool that will be very useful a day later when the footing forms are set. The tool can be borrowed, bought, or rented. The 'dozer driver might even have one.

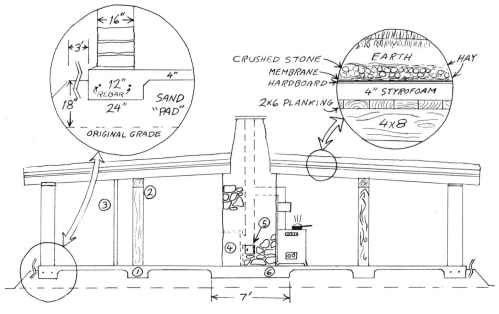

NORTH-SOUTH CROSS-SECTIONAL VIEW

Fig. 89. Cross-sectional view of the Round House, along the north-south axis.

The alert reader will notice that a radius of 23′ was used in the calculations, but only 22′ was measured from the center point to describe the circle. Figure 90 shows that the slope of the outer edge of the pad returns the volume to approximately the calculated figure. In any event, the calculation for the in-bringing of material should only be considered an approximation. Compacting, surface irregularities, and inaccurate measuring by the haulage contractor may combine to send the calculated figure off by 10 percent or more. The important consideration is that there be a minimum of 3′ of flat berm all around the footing (the berm is the cross-hatched area on the diagram).

The whole pad thus created should be compacted with a powered compactor, available at tool rental stores. Compaction is best accomplished on damp material. The material should be compacted in 6″ layers.

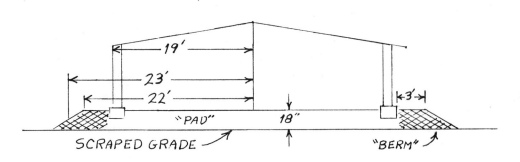

Fig. 90. Pad for the Round House, vertical scale exaggerated.

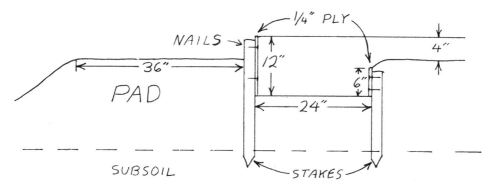

Fig. 91. Cross-section of footing shuttering.

THE BULL'S RING. A pipe with an outside diameter of 1½″ and with a threaded end should now be driven into the pad at the exact center, replacing the stake used previously. The pipe should be driven carefully so that the threads are not damaged, and should project about 8″ above the pad. To get accurate radial dimensions, I recommend the purchase of a bull's nose ring or similar device that will fit easily over the pipe, a line with low stretch characteristics, and a plumb bob. Join the plumb bob and the bull's ring with the line and place the bull's ring over the pipe. The distance from the center point of the house to the edge of the footing should be 4″ greater than the radius of the house. In our example the radius is 19′, so the footing form should be set at 19′4″.

THE FOOTING FORMS are made of ¼″ plywood, which is tough and will bend easily to such a gradual curve. A sheet of plywood will yield four strips 12″ wide by 8′ long, or 32′ of forming material or *shuttering*. The outer circumference of the footing equals 38′8″ times *pi* (3.1416), or 121.5 feet. Four sheets of plywood, then, will yield enough material (128 lineal feet) for the outer shuttering.

The inner shuttering for the footing is made with 6″ strips of ¼″ plywood. The cheapest grade of plywood is good enough for shuttering material. Figure 91 shows a cross-section of the footing with the placement of the stakes for supporting the plywood. Figure 92 shows a top view of the shuttering, and the placement of three stakes per section of plywood. The larger 2-by-6 stakes serve

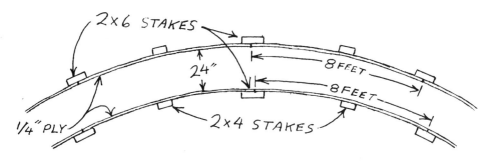

Fig. 92. Top view of footing shuttering.

double duty, supporting the ends of two adjacent sections. The top of the footing should be 4″ above the mean grade of the pad. The stakes and the 8″ of earth-supported shuttering below pad grade is sufficient to keep the forms from bowing outward. The sand on the outside of the shuttering should be tamped to assist in this supporting.

The inner shuttering for the footings and the supporting stakes is left in the ground permanently, as the floor and footings are poured at the same time. With a diameter greater than 40′, it would be necessary to pour the footings and floor separately, with an expansion joint between, as shown in Figure 70 on page 77.

THE OCTAGON

There are eight posts in the house which carry an octangular inner ring beam to support the long rafters halfway for increased roof load capacity. The house is designed to carry a 6″ earthen roof with a 4′ snow load, or 120 pounds per square foot. In northern New York, most houses are designed to carry 40 pounds per square foot. The eight posts, then, each support a heavy load, and require individual footings 1′ deep (including the 4″ floor) and 2′ square. The shuttering for these pillar footings can be built from one sheet of plywood, divided and nailed together as shown in Figures 93–95. All of the 2-by-2 corner pieces can be made from one 8′ 2-by-4, ripped down the middle. The chamfered edge shown in the lower diagram lessens the possibility of the concrete shearing between floor and footing.

THE HEAT SINK

The massive fieldstone heat sink (5′ in diameter) at the center of the structure also requires a footing. This footing is round, 7′ in diameter, and the same depth (12″) as the other footings. Again, ¼″ plywood cut into 6″ strips will serve as the shuttering, which is left under the floor.

Note: Because the heat sink is load-supporting, the eight posts (and their footings) are not positioned halfway from the center point to the perimeter. They are placed so that the two spans of the primary rafters are equal. The center of the pillar footings are actually 10′4″ from the center of the house. The location of pillar footings should be accurately transposed from the plans, using known reference points. A good idea is to mark the four primary compass points with stakes, and later, before the concrete has set, with heavy scratch marks on the perimeter. These marks will be very useful later on in locating the positions for the posts, as the floor will obscure all trace of the footings.

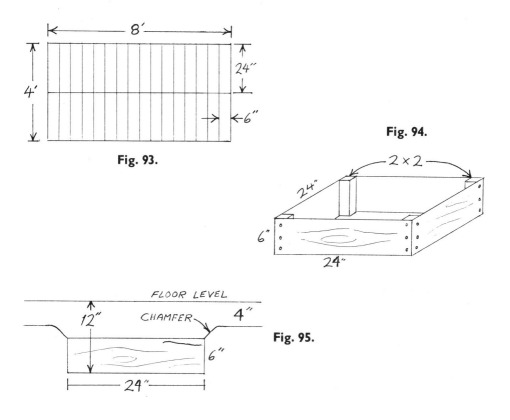

Fig. 93.

Fig. 94.

Fig. 95.

Figs. 93, 94, 95. Shuttering for pillar footings.

The outermost footing forms should be levelled at 4″ above the pad level. All interior shuttering, which will be left in the pour, should be set to a level 6″ below the outer forms, or approximately 2″ below the grade of the pad. Pouring the pillar and heat sink footings first will decrease the likelihood of sand spilling into these.

The rest of the preparations for the pour consist of laying a double ring of ½″ rebar around the perimeter (Figure 89); placing wired rebar in the pillar and heat sink footings; installing all underfloor plumbing, such as drainage pipes and the incoming line from the well; and unrolling wire mesh floor reinforcing. All exposed piping, including the center pipe, should be covered to protect against damage or filling up with concrete. I use rags and plastic bags for this.

Two recommended options are to place 1″ of Styrofoam® on the pad where the floor is uninterrupted by footings, and to run a 4″-diameter underfloor air vent to the stove locations. The air vent supplies oxygen for combustion and eliminates most drafts caused by the stoves trying to suck in air from around windows and doors. The outside end of the vent pipe should be protected from insects, rodents, and water penetration. If the Styrofoam® is used, grades should be adjusted to assure a full 4″ floor. The Styrofoam® allows the concrete floor to operate at a higher temperature as a heat sink.

95

THE POUR

You should use three-thousand-pound concrete. The total volume of the pour is 22¼ cu yd. An order of 23 cu yd should be safe; or, pour two 9- or 10-yard truckloads and estimate the remaining concrete required. The concrete should be poured as stiff as is manageable. Soupy concrete is much easier to pour, but there is a much greater risk of shrinkage cracks, especially where the floor meets the various footings. There should be one member of the crew whose only job is to make sure that the reinforcing is held in the middle of the pour, both rebar and mesh. The reinforcing can be lifted with the tines of a rake. It is particularly important that the mesh not be near the surface, especially if a powered trowel is to be used. The spinning blades of the power trowel might get caught in the mesh, which will pull the mesh out of the concrete and put the trowel operator in some jeopardy. I recommend the use of a power trowel to obtain a smooth flat floor, but I also suggest that the tool be used by an experienced operator, even if it means hiring someone for a few hours.

I have gone into some detail on this foundation method, as information on circular foundations is not readily available, and also because the foundation is arguably the most important structural component in any house. A more detailed discussion of the techniques involved in pouring footings and floors is found in my other book, *Underground Houses* (Sterling, 1979).

When the floor is dry and the outer forms are removed, the location of the intersections of the interior rooms with the cordwood wall should be marked on the perimeter with a bright crayon. This enables the builder to locate the doors and windows during wall construction. The two door frames will have to be built, plumbed, and temporarily braced. A pleasing alternative to rectangular door openings is the arched door opening suggested by Jack Henstridge and incorporated into the Sam Felts home in Adel, Georgia (page 115).

CORDWOOD

With the door frames in place, cordwood construction is now ready to begin. On the house we are building, the northern hemisphere will be constructed of 75-year-old split cedar fence rails, the southern hemisphere of a variety of split hardwoods. As all log-ends are 16″ long, this is the best use of the material with regard to the R-factor of the woods involved. By this method, the northern (and colder) hemisphere will have an insulation factor about 30 percent greater than that of the southern hemisphere.

Laying up log-ends is extremely simple in the round house. First, an 8′ pipe is screwed onto the threaded pipe which should be sticking up out of the floor by 4″ at the exact center of the house. The pipe should be plumbed by three wires or ropes tied to heavy stones or blocks placed about 10′ from the center (Figure 96). A stronger alternative would be to fasten it to eyebolts set in the concrete pour specifically for this purpose. Later, they can be cut off with a hacksaw. It is important that this pole be exactly plumb; therefore, it should be checked with a level or a plumb bob every day before work.

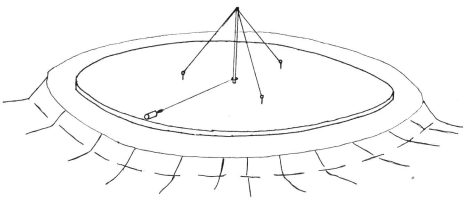

Fig. 96. With the center pipe plumb, laying up log-ends to the line assembly is a breeze.

The first course should be laid up in an uninsulated mortar joint a full inch thick. This allows a slope away from the cordwood wall to be incorporated as a part of the footing. This slope, which can be seen in the footing inset of Figure 89, prevents the collection of water against the base of the cordwood wall. A further precaution for the first course is to use a wood that has good resistance to rot, such as cedar. Do not use poplar on the first two or three courses.

Now that the center pole is plumb, laying up log-ends is a breeze. Set the length of the bull's-ring/plumb-bob assembly so that it is exactly 17′8″ from the very center of the pipe to the point of the plumb bob. The 16″ log-end is laid so that its interior surface just touches the plumb-bob assembly. The proof of the pudding is to check the width of the footing outside of the wall: if the log-end is a true 16″, there should be 4″ of footing showing.

By continuing with this method, the wall will be exactly round and the interior surface exactly flush, which is as it should be. The exterior surface will feature more relief, due to the slight variations in log-end lengths that are bound to occur. The procedure is to start at one door frame and to lay a single course of log-ends all around the perimeter until the work returns to the other side of the door frame. Then a double bed of mortar, each 4″ to 5″ wide, should be laid with a small trowel. Insulation (fiberglass, sawdust and lime, shredded beadboard, etc.) is installed in the cavity between mortar beds. Then, the next log-end is put in place. The inner mortar joint should be between ¾″ and 1″ thick, the same as for the built-up corners method. The outer mortar joint will be wider in the horizontal direction, as previously described.

For easy reference, I will repeat the recommended mortar mix for load-supporting cordwood masonry: 3 sand, 3 sawdust, 1 Portland cement, 1 lime. The equivalent mix using masonry cement is: 6 sand, 6 sawdust, 3 masonry, 1 lime.

When the wall is up about 8″ on average, it is time to jack up the string to the 12″ height on the pole. Otherwise, the wall will start to curve inward and you'll get a cordwood igloo instead of the round house you want. If the bull's ring is jacked up as the wall height increases, you'll end up with a perfectly plumb wall. There are various methods of holding the ring up at the desired height. One is to drill a ³⁄₁₆″ hole through the pipe every 12″. A sixteen-penny nail can be used as a pin to keep the ring from slipping down. Another method is to use a nylon expansion ring with a 1½″ internal diameter. These rings are used in plumbing expansion joints to stop leaks. They will not slip on the pipe and, if the outside diameter of the ring is great enough, they will stop the bull's ring from sliding down the pole. You may discover your own method of keeping the bull's ring at the correct height.

Window frames should be pre-built, as shown in Figure 73 (page 80). When the masonry has progressed almost to window height, an effort should be made to flatten the top of the cordwood stack for the placement of the window frame. The edge of the frame, where contact with the mortar joint will be made, should have a number of roofing nails driven to within ½″ of home, so that a better frame-to-mortar bond will result. The frame should be temporarily braced until it is completely supported by the wall, as described in the previous chapter.

Attention should be given to some of the special design features that are possible with cordwood masonry, such as shelves, "bottle-ends," coat-pegs, patterns, and the like. These are discussed in Chapter 7.

Remember to cover the top of the work with plastic at the end of each day, to protect the wall insulation from rain damage and the log-ends from swelling.

Lintels should be placed directly above the door and window frames. Dry 6-by-6 landscaping timbers are the best material for the job, although logs milled on two sides would also be fine. Lintels should be supported by at least 10″ of cordwood masonry on each side of the frame. Insulate between the inner lintel and the outer lintel. Do not use creosoted landscaping timbers as interior lintels.

The plate on the round house serves only to distribute the rafter load equally over the wall. It is not needed to tie the wall together or to prevent outward pressure on the wall, as will be seen later. I recommend an inner and an outer ring of 2-by-6 plates. The inner ring will be made up of 32 pieces, each 3′5″ (41″) long. The outer ring will also have 32 pieces, but they will be 3′7″ (43″) long. Two of each can be cut from a 14′ 2-by-6, so you'll need sixteen 14-footers for the plate system, 224 board feet in all.

If a standard 80″ door, door frame, lintels, and a 2″-thick ring plate are all to be used, preparation for the laying of plates will be made before the wall reaches a height of 7′4″ (88″) (Figure 97). My personal preference is to

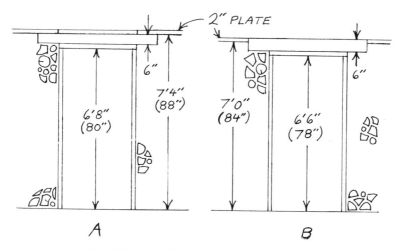

Fig. 97. Two different door installations.

go with a 6′6″ (78″) door and to have the door lintel incorporated into the ring plate as shown in the second example. This saves about 5 percent of the wall labor and cordwood required, and keeps the ceiling a bit lower, which is pleasing with an exposed rafter system and makes the house a little easier to heat as well. Remember that the perimeter is the lowest point of the ceiling, which begins immediately to slope upward at a very gradual pitch of about 1:18. By the first method the actual height to the planking would be 9′2″ at the heat sink, as opposed to 8′10″ by the second method. With the harsh reality of energy costs, the days of the very high ceiling, at least in new structures, are over.

No log-end should be laid so that its highest point is within 1″ of the height at which the plates will be laid. After the last course of log-ends is laid, a thick double bed of mortar should be laid so that the plates can be levelled easily at a common top-edge height of 7′6″ (or 7′2″, if the second door-framing method is employed). The levelling of the plates is quite important and it would be a good idea to use a transit or surveyor's level for the job.

The placement of the plates is also important in that it determines the location of the individual rafters. This house is designed so that there is a rafter at each of the 32 points found on a ship's compass card (Figure 98). The easiest way to find the exact placement of the plates, then, is to have already marked these compass points accurately on the edge of the concrete footing with an orange crayon (for example) and then transpose the marks to the top of the wall with an 8′ straight stick and a level, or a plumb bob.

The 64 plate pieces should all have their midpoints marked with a pencil line and these midpoints should be aligned with the compass points. There will be about ½″ between the ends of adjacent plate pieces, which can be filled with sawdust mortar. Old nails left sticking out ½″ from the underside of the plate pieces will help to form a bond between the plate and the masonry.

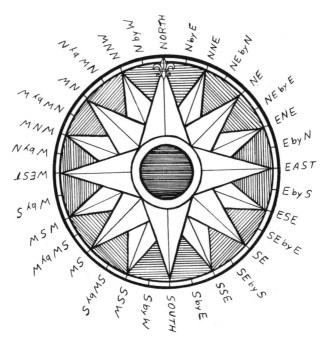

Fig. 98. A ship's compass card.

We should not forget to cover the top of the wall before moving on to the inner support structure.

It's a bit of a toss-up as to whether the fieldstone heat sink or the octagonal support structure should be built next. If individual circumstances point towards building the heat sink first, be sure to mark the locations of the eight posts before you lose the advantage of having an exact-center reference point.

THE OCTAGONAL FRAMEWORK

The house that I'm building features old 8-by-8 barn timbers of spruce and oak. I feel that these old hand-hewn beams complement the cordwood very nicely and help to create a warm, rustic atmosphere. The posts should be cut to a length so that the addition of the 8"-thick beams will establish a total height for the octagon that is 6" higher than the top of the wall plate. Posts, then, need to be 7'4" or 7'0" high, depending on which of the two methods of door framing is chosen. If an earthen roof is not desired, I recommend a steeper pitch to the roof and a reduction in rafter size from 4-by-8's to 2-by-8's.

The beams of the octagon can be spiked into the posts, but a stronger and more satisfying procedure is to join the two beams to the post with a mortise-and-tenon joint. This will be challenging even for an experienced carpenter, because of the angles involved.

100

THE STONE HEAT SINK

The round stone thermal mass at the center of this house serves double duty as load support and heat sink. If it is built 5′ in diameter, the column of stone will weigh about 15 tons. The need for heavy compacting directly beneath the heat sink footing should now be clear.

Again, the central pipe can be employed to keep the mass plumb and round. It can be re-guyed to the wall plate. The outer surface of the wall should be built carefully, paying attention to basic masonry principles, such as keeping a consistent width of mortar joint and not placing one vertical joint directly above another. For stone work, I am very happy with a mortar mix of 5 sand, 1 Portland, 1 masonry. The interior of the stone cylinder can be filled with rubble and mortar. Solid masonry should not extend all the way to the flue tile. A 3″ space next to the flue tile should be filled with a weak (10 to 1) mixture of sand and lime, to allow for expansion of the flue tile during intense heat, as will occur during a chimney fire. (Chimney fires can be avoided, of course, by adopting intelligent wood-burning practices, such as using dry wood and periodically checking and sweeping the chimney, but a house should be designed to contain them should they occur — and they surely will, sooner or later.)

A cleanout with a tight-fitting door should be installed at the base of the flue. Ceramic thimbles are used to change over from stovepipe to masonry. Again, the thimble should be set in the sand-and-lime mixture to allow it to expand.

My plan is to set an airtight cooking range against one side of the mass, and to set a parlor stove into the mass on the opposite side. Setting the stove into the mass helps to charge the heat sink more efficiently; also, it keeps the stove farther back from the seating arrangement of the living room, protecting guests from overheating and providing more room for movement. I recommend an airtight parlor stove which can have the front doors opened to allow viewing of the open fire. I do not recommend a fireplace, as they are grossly inefficient and suck the home's heat right up the chimney. Ambitious and crafty types may want to tackle a masonry stove, which is extremely efficient. These stoves are common in eastern Europe and are making a bit of a comeback in this country.

It is my intention to discuss those aspects of building directly connected with wood masonry, not stone masonry (about which whole books have been written). I will make three brief points, however, pertinent to this particular design recommendation. One is that the connection of two stoves to one flue is not recommended by some advocates of wood burning. Personally, I've had very good luck with connecting two stoves to one chimney. The key is that one stove should enter the chimney at least 3′ higher than the other. Another good piece of advice when burning just one stove is to have the other stovepipe closed off with an *airtight* damper.

The second point is that some building codes and inspectors frown on supporting rafters with a mass of masonry in direct contact with a chimney. My response is that the 15-ton mass of this design is so great that the heat sink operates at a very low temperature, especially at the 9′ level. Furthermore, there are a full 20″ between the rafters and the flue tile. My advice is to build the chimney as shown in Figure 99, using five chimney blocks and a 3″ fiberglass-filled cavity outside of the chimney blocks. This way, the concrete support shelf for the rafters will not get hot enough to crisp the wood, even during a chimney fire. "Crisping" means weakening the wood fibers by a rapid over-drying of the wood. Many master woodworkers believe that kiln-drying "crisps wood to death."

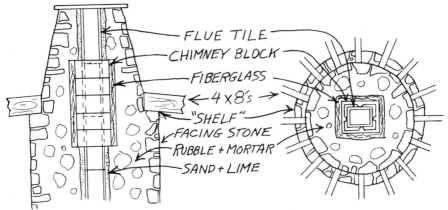

Fig. 99. Chimney construction.

The third special consideration for this design concerns the fastening of the 16 primary roof rafters to the stone mass. You will note that this method also fastens the rafters *to each other*, eliminating outward pressure on the cordwood walls during construction of the roof. A ½″ steel rod or ½″ rebar 14″8′ long is bent into a ring and the ends are welded together. A ¾″ steel rod or rebar 10′ long is cut into 16 equal pins of 7½″ each. Each pin is welded to the inside of the ring, equidistant from each other. Three-and-a-half inches of each pin will show above the ring, 3½″ below. The pins are welded perpendicular to the ring. Their center-to-center spacing is exactly 11″ (see Fig. 100). This ring can be made quite easily by anyone using a welder. It is set into the masonry shelf where the 16 primary rafters come together. The lower part of the pin anchors the unit to the stone mass; the upper part will receive the 4-by-8 rafter. The rafters should have holes 1″ in diameter drilled into their undersides to a depth of 4″. These receiving-holes should be slightly angled to take into account the roof pitch, as shown in the diagram.

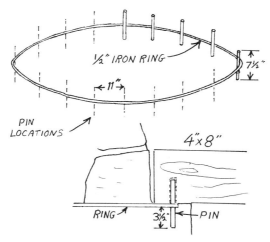

Fig. 100. A ring assembly for tying rafters to the central stone mass.

THE RAFTER SYSTEM

There are 16 primary and 16 secondary rafters, seen clearly in Figure 87. All rafters are 4-by-8 spruce, hemlock, or some equally strong wood. The primary rafters are 18′6″; the secondary rafters are 12′0″. The long, single-piece primary rafters offset the sag which would occur if they had been constructed of two pieces joined over the octagonal framework (Figure 101).

Fig. 101. One rafter covering two 8-foot spans offsets sag and is much stronger than two rafters spanning 8 feet each.

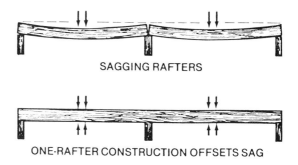

SAGGING RAFTERS

ONE-RAFTER CONSTRUCTION OFFSETS SAG

The extended cantilevers of the secondary rafter system work in a similar fashion. Setting the primary rafters is a three-man job and is made easier by first constructing a template of 1-by material showing the positions and shapes of the *birdsmouths* (see Fig. 102). Birdsmouths are checks on the underside of the rafter designed to allow the rafter to sit flat on the support structure. The template must have two different center settings as the central birdsmouth is located at different positions, depending on whether the rafter passes over the points of the octagon or over the midpoints of its sides. The shorter secondary rafters require a different template. The secondary rafters must be installed so that they are in the same plane as their two primary rafter neighbors, so that planking can be nailed on properly.

103

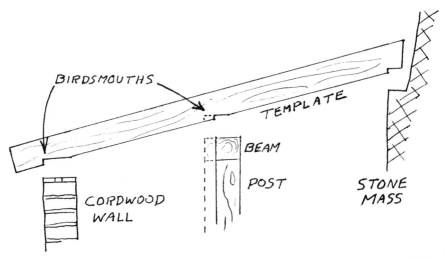

Fig. 102. In setting the primary rafters, first construct a template of 1-by material showing the position and shapes of the birdsmouths.

PLANKING

The roof planks are 2-by-6's. Dry tongue-and-grooved material is best, but rough-cut 2-by-6's that have been planed on one side for uniform thickness will also serve well. Green wood should be avoided, but if there is no other choice, the green planks should be nailed as shown in Figure 103. Planking should commence at the edge of the rafter system and proceed towards the middle. The cut-offs are used as the triangle gets smaller, so there is little waste. Including plate material and allowing for waste, approximately 3,000 board feet of 2-by-6's are required to build this house.

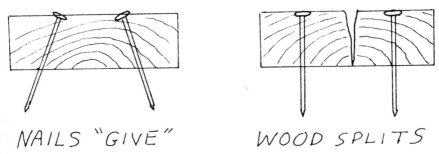

Fig. 103. Angling the nails makes it easier for the wood to shrink without splitting.

The roof is insulated with between 2″ and 4″ of Styrofoam®, depending on the climate. Strictly speaking, the word "Styrofoam" refers to Dow Chemical's extruded polystyrene product and is trademarked by them as such. It is one of those words like "Band-Aid" and "Coke" that are often mistaken for vernacular English. "Styrofoam" cups are not made of Styrofoam. Beadboard and polyurethane foam are often mistakenly called "Styrofoam." The difference is very important. Extruded polystyrene has an R-factor of about 5.25 per inch and is *close-celled,* making it much more resistant to deterioration from moisture. Beadboard is quite a bit cheaper but has an insulation value of about R3.5 per inch and is more subject to deterioration. Polyurethane foam has a high value (R8.5 per inch) as insulation, but loses most of its insulative qualities if exposed to moisture. And if it catches on fire, the gases are deadly.

Figure 104 shows how 4-by-8 sheets of rigid foam insulation can be cut and fitted on the roof triangles with virtually no waste. The roof has 16 triangular facets. The facet labeled *8* on Figure 87 shows the placement of the rigid foam insulation. Each of these facets requires two entire 4-by-8 sheets, cut as shown in Figure 104, Pattern One, and ¼ of a sheet cut as shown in Pattern Two. Thus, a sheet cut according to Pattern Two will serve four facets. Thirty-six sheets of insulation are required to cover the roof surface and they are secured with a few nails and washers until the hardboard layer is applied.

THE HARDBOARD LAYER provides a hard, flat surface for the application of the waterproof membrane. It serves no other purpose, although I suppose it

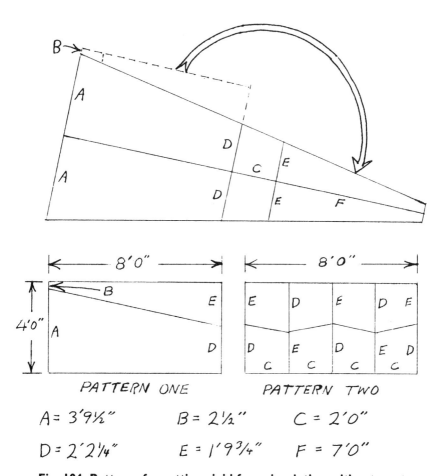

PATTERN ONE   PATTERN TWO

A = 3′9½″   B = 2½″   C = 2′0″

D = 2′2¼″   E = 1′9¾″   F = 7′0″

Fig. 104. Patterns for cutting rigid foam insulation without wastage.

adds some insulative value to the roof. At Log End Cave I used ⅝″ particle board, mainly because I got a good deal on it, but ½″ would have done just as well. Plywood would have done too, of course, but it is generally a lot more expensive than particle board, unless it is salvage. The sheets are fastened to the planking with spikes, six to a sheet. Drill nail holes through the particle board first or the sheets will break. I keep the spikes about 4″ from the edge of the sheets. Depending on the thickness of insulation used, a 2-by-4 or 4-by-4 is nailed all around the overhang perimeter as shown in Figure 87. The hardboard can be fastened to this perimeter board with 8-penny nails. Fitting the hardboard without waste is demonstrated in Figure 105. In this case, Pattern Two yields enough for the center pieces of two adjacent facets, not four as with the insulation pattern. Forty sheets of hardboard, then, are required to cover the roof.

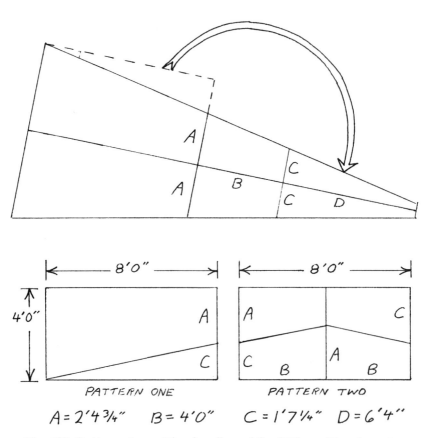

Fig. 105. Patterns for cutting hardboard for fitting without waste.

WATERPROOF MEMBRANE

There are many different ways of waterproofing the roof. On our underground house, I used 6-mil polyethylene and plastic roof cement, as described previously in the sauna construction (pages 65–72), and in more detail in *Underground Houses.* Another possibility is a built-up roof of three layers of 60-pound roofing, with hot pitched tar in between. Several companies are bringing out various liquid polymer "waterproofing in a can" products. These products bear further investigation. It is outside the scope of this book to enter into a thorough discussion of waterproof membranes. The question is discussed in several books now making their appearance on earth-sheltered housing, and I would suggest that the reader make his own investigation into the various alternatives before proceeding.

Flashing and retaining ring systems should be similar to those described for the sauna in Chapter 3.

THE EARTHEN ROOF

The inset in Figure 89 shows the various layers of the earthen roof above the membrane. First, a full 3″ of hay will protect the membrane during the application of the second layer, which is crushed stone. Another 3″ layer of hay covers the stone. Eventually, these layers of hay will decompose to form a thin, fibrous mat that protects the membrane and keeps dirt out of the stone, which is used for drainage. Five or six inches of topsoil are placed on top of the hay and planted with grass seed — mixed timothy and rye is good — and flower bulbs such as crocus and snowdrops which do not require deep soil. A ½″ layer of hay mulch will help retain moisture while the grass gets established. This roof is not meant to be mowed. The shaggy grass will give a warm pleasing effect not unlike thatch, and the roof will retain moisture better through the dry seasons.

INTERNAL WALLS

The use of 2-by-3 framing (full sized, once through the planer) is recommended for internal walls. This material allows the application of ½″ plasterboard on both sides of the wall and the plasterboard will be flush with the edges of the roof rafters. It is a great time-saver and makes for a much more attractive wall if the internal walls are constructed directly beneath a beam or

Fig. 106. Log-ends fill the space between rafters from the top of the beam to the roof planking and go well with the rough-hewn style of the house.

a rafter. Just missing a beam or a rafter with a wall is sloppy and trouble-some, besides providing a place for cobwebs to develop. Only in the living room does an internal wall not rise directly underneath a beam; this is to give more room and clearance for the parlor stove. As there are no log-ends in evidence in the main living area, it might be a pleasing design feature to construct this wall of 5″ or 6″ log-ends, using a full-width mortar joint. Also, above the octagonal ring beam, short log-ends can be used to fill in the spaces between rafters. We did this very successfully at Log End Cave (Figure 106).

ELECTRICAL

Electrical wiring can be hidden in the internal walls, buried under the slab in conduit, or threaded through the cordwood, also by the use of conduit. This latter method was adopted successfully by Malcolm Miller in his home in New Brunswick (pages 137–138).

THE FLOOR PLAN

The suggested floor plan (Figure 88) for this round house is an adaptation of the plan used at Log End Cave. It is as if the Cave floor plan had been stretched around a curve. Entry is through a mudroom, which serves triple duty as a cloak-and-boot room, thermal break, and fuel store. The kitchen/ dining/living area is open plan. The kitchen area features two large double-pane windows facing south for a view and the maximum utilization of solar energy. The pantry is conveniently adjacent, as is the dining area. In our underground house, the pantry also serves as the battery storage area for our wind electric system. A room divider defines the library or play area. Skylights are placed over the dining, living, and library areas. It is said that a skylight admits five times more light per square foot of glazing than a wall window. Visitors to our home have been surprised at the amount of light admitted by the skylights.

NOOKS. One of the special design features of the round house is the use of two separate *nooks*. These nooks serve several purposes: their presence means that there are no doors opening directly into the dining, living, or library areas, maximizing the usefulness of the wall space. The morning nook — so called because it faces the summer sunrise — is a convenient location for the home's second entrance. The nooks are 6′ by 8′, large enough for a sewing area, a desk, or just a private reading area. The nooks let light into the main room and show off some of the cordwood masonry to the living area. An attractive addition to these nooks would be the stained glass windows, accenting blues, greens, and yellows in the morning nook, and reds, oranges, and browns in the afternoon nook.

COMPASS. I have said that the rafters correspond to the 32 points on a compass card. Each rafter in the main living area could be labeled in some way with the correct compass point. The house would then be an accurate direction finder, on a par with Stonehenge. There are times when this could be useful, such as for finding planets and satellites. If this feature is important to the builder, care should be taken to orient the placement of the eight pillar footing forms in reference to Polaris, the North Star.

HEATING AND COOLING

This house has several heating and cooling advantages. Its round shape presents a much easier object for cold winds to move around — sort of like a mushroom, really, and unlike the "sail" that most houses with large flat walls present to the wind. Its centrally located radiant heat source transmits heat equally in all directions, and there are no square corners far from the heat source. The tremendous thermal mass in the concrete foundation and floor, the fieldstone heat sink, and in the cordwood masonry itself, helps to maintain a steady temperature in the home, effectively averaging the extremes of temperature which occur over a period of time. Finally, the earthen roof aids in summertime cooling by eliminating the hot asphalt roof, and also through the cooling effect of water evaporation. The floor plan of the house is designed so that the living and kitchen areas can be maintained at 70° or 72° F.

(21°–22° C.), while the peripheral rooms will naturally keep about 5° to 8° F. (3°–5° C.) cooler, because of the dead air space of the internal walls. Jaki and I — and most people we discuss this matter with, including doctors — agree that a bedroom that is cooler than 70° F. (21° C.) is also healthier. My best estimate is that three cords of wood per year will be required to heat this house in upstate New York, most of it burned in the cookstove.

Note: In 1981, we began construction of "Earthwood," a house similar to the Round House on pages 88–89, except that it is a two-story design. Our plan called for earth-berming the northern part of the house for greater heating and cooling characteristics, so we decided to use dry split hardwood on the earth-sheltered portion of the wall (because of hardwood's greater thermal mass) and old fence rail log-ends made of split cedar on the southern hemisphere for their greater insulation value. The hardwood caused us a problem, *because it was actually too dry* (split and stacked three years).

The house is built on a floating slab. During construction, rainwater would collect against the base of the wall on the interior, causing the superdry hardwood to swell and the wall to begin to hinge upwards and outwards on the outer mortar joint. When a 6′ high portion of the 16″-thick wall became 3″ out of plumb, we decided to dismantle the otherwise beautiful hardwood wall. The cedar part of the wall, *even though it experienced the same conditions,* did not swell and tip outward. The lighter, airier cedar seems to act as its own "expansion joint." The wood fibers are able to swell into the log-ends' own air spaces without putting undo pressure on the mortar joint.

As we did not have enough cedar to build the entire house of cordwood, we decided to construct the below-grade portion out of 8″ by 16″ concrete corner blocks dry-stacked widthwise in the wall and surface-bonded. The W.R. Grace Bituthene® waterproofing membrane and 2″ of Dow Styrofoam® applied to the exterior enables the concrete to act as a very effective thermal mass.

A correspondent in Illinois has experienced a similar kind of problem with dry pine, as had Andy Boyd with dry hardwood. (See page 129 and Color Plate H3.) Andy found that swirling his dry hardwood for a few seconds in a five-gallon pail of water prior to laying up alleviated the problem. The lesson is that, contrary to my previous experience and knowledge, it is possible to use wood which is *too* dry, although white cedar seems to have a built-in protection against swelling problems. If in doubt, the builder should make a 3′-high by 6′-wide test panel and subject it to the kind of damp conditions expected during house construction. Allow at least a week for signs of stress-cracking in the mortar joint.

For a further discussion of the Earthwood design, see *Money-Saving Strategies for the Owner/Builder* by Robert L. Roy (Sterling, 1981).

6. Cordwood Homes and Then Some

Wood masons seem to have an almost childlike enthusiasm for their work, and will tell anyone who comes along as much about their cordwood technique as they can. This probably derives from the memory of their own thirst for information at the time they were ready to begin their own houses; and, too, cordwood seems to appeal to a particular sensibility, one which is open, cheerful, and uninhibited. Or is it that cordwood influences people that way after the fact? I'm not entirely sure, but, happily, there is one generality that seems to apply to all the cordwood people I have met, and that is that they are glad to help "spread the Word" and add to the literature by sharing their experiences. I sent questionnaires to several owners of cordwood homes who were known to me. Not one took advantage of the option to remain anonymous. These people are rightfully proud of what they have done and are eager to help others accomplish similar goals.

The structures described in this chapter range from a $180 hut to a 3700 sq ft house worth well over $100,000 and built in an approved subdivision with bank financing. Nine of the 15 examples cost less than $9,000 to build, inclusive of materials for the basic house and all paid labor. The figures quoted do not include the cost of land, well, septic system, or home furnishings, unless otherwise noted, and are, of course, subject to inflation as time passes. Canadian home costs are given in Canadian dollars.

The House on Page 83

(Figure 107)

In Jack Henstridge's book, *Building the Cordwood Home,* he included several suggested plans for curved-wall homes. He'd never built any of the homes himself, but knew that they were workable plans. One day in 1979 he received a phone call from Bernie Curran, of Chesterville, Ontario, hundreds

Fig. 107. The "House on Page 83."

2 plus bedrooms

Blown acrylic dome for skylight over kitchen. Using a 9" block — less than 5 cords of wood are required to build the exterior wall. The heavy lines on the interior are 6" blocks. For structural strength, only the block wall around kitchen area is necessary. The others are just for effect. Less than 3 cords of wood are required.

Total: 8 cords peeled pulpwood @ $40.00 = $320.00, plus lime, cement, and sand.

In the winter, install a door flush with exterior wall at entry. This would give you an air-lock effect.

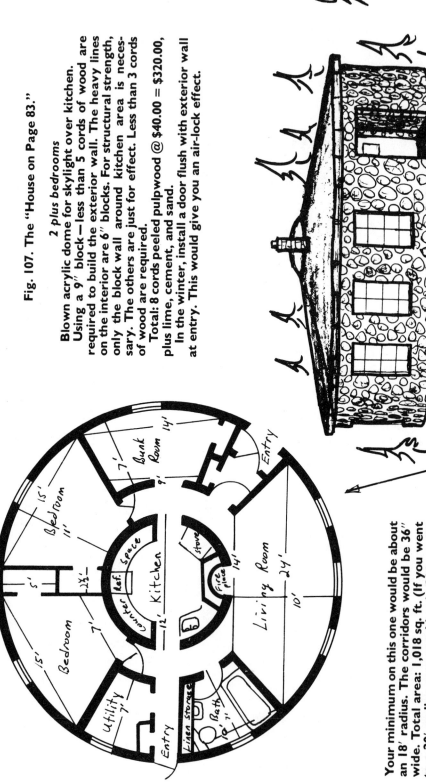

Your minimum on this one would be about an 18' radius. The corridors would be 36" wide. Total area: 1,018 sq. ft. (If you went to a 20' radius, you would gain 239 sq. ft.) Measurements here are based on an 18' radius.

of miles west of Jack's New Brunswick home. The conversation went something like this.

Bernie: "Hello, is this Jack Henstridge?"

Jack: "Unless I owe you money."

Bernie: "Jack, I just built the house on page 83."

Jack: "You did *what?*"

Bernie: "I built the house on page 83."

Jack: "I'll be right there!"

It must have been an emotional experience for Jack to visit the Currans and see his idea made a reality. With Jack's permission, I include a reprint of the inspirational page.

The previously inexperienced Bernie Curran, working with contractor and friend Bill Hoople, built the house with 12″ cedar log-ends cut from old utility poles. It took them five months. Even though the mortar they used was quite strong (6 sand, 3 masonry, 1 Portland), it did not shrink. This may be because Bernie used thick mortar joints, between 1″ and 3″. Relatively large volumes of mortar like this may have retained moisture longer, prolonged the set, and reduced the shrinkage that is normally found in a strong mix.

Equally surprising, their log-ends have shrunk a bit and have had the worst of their checks "plugged with oakum." Bernie says he will be caulking later. I think this shows the need for drying even old utility poles for a couple of months under good drying conditions *at the ultimate log-end length,* because the internal stresses of the wood change as the long pole is cut into 12″ pieces. Splitting large log-ends is suggested. Another consideration, pointed out by Andy Boyd of Arkansas, is that log-ends laid up at a very humid time of the year may be temporarily swollen with moisture, even though they had previously achieved ambient moisture content. Likewise, dry log-ends that have been out in the rain are no longer dry and have to be re-dried. So keep those wood piles covered!

The Currans included several personal design features in their home, such as a kind of a wagon-wheel rafter support beneath the Plexiglas domed skylight, brick support columns in the kitchen, old barn-board walls, and patterned stucco walls. Their 1,150 sq ft home cost them $8,515.21 — somebody kept track — and another $6,000 in labor. The Currans obtained a bank loan on the project. Based on the 17 percent exchange rate of 1979, the total cost of the Currans' home in U.S. dollars was about $12,000, or $10/sq ft, including the hired help. The expense figures in these "histories" may startle some readers, but they are for real. Low cost is one of the great advantages of cordwood construction. Mrs. Curran tells of some of the others:

"We really enjoy living in our round log house and wouldn't want to live in conventional housing. One friend described his feelings about our house by saying he felt he was in 'the gingerbread house.' There's a warm, cosy, fun feeling when you walk into our home. Our heating costs are nil, a very nice economical feature. We heat with a Fisher wood stove and have a free supply of wood."

Fig. 108. The Felts Home.

2 bedroom (15' radius — 707 sq. ft.)

14' diameter blown acrylic dome skylight. It is expensive, but total cost of the building is still very low. A ⅜ geodesic dome would save a lot of money.

There is a ladder up the block wall around the indoor garden to a loft built over the corridor (about 3' wide). It can be used for an extra sleeping area, or just a place to go and gaze at the stars.

The arches inside would give the place the feeling of an old castle, and the curved walls give the feeling of a much larger size simply because you can't see all that far. The sunken tub is 3' wide and 6' long and as deep as you want to make it.

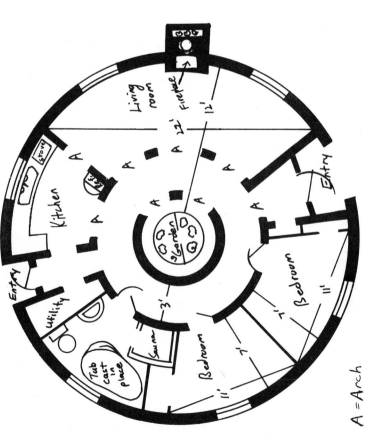

A = Arch

To make "arches," cut ¼" plywood in strips equal to width of wall. Bend plywood to 1" board for spreaders and prop in place. Build the wall around it. When "cured," remove the form. The wall in this drawing is 12" thick.

The Felts Home

(Figure 108)

If the Currans' home is "the house on page 83," then Sam Felts's place is arguably "the house on page 82" of Jack's book. Actually, Sam has modified Jack's design quite a bit by eliminating the center circle and adding a 13'-diameter rotunda, which has a 17'-high ceiling.

As the house is built in Adel, Georgia, the 9"-thick red cedar log-ends used on the exterior wall are more than adequate for insulation. Sam used 6" wild cherry log-ends for interior walls. The cordwood was air-dried for 11 months. The log-ends did shrink some, however, and the worst of them were caulked with Z-brick cement. Sam's mortar mix of 8 sand, 6 sawdust, 3 Portland, and 2 lime did not shrink. The house is large: 40'9" external diameter, 1300 sq ft. The cost was $15,000 in materials and $6,000 for labor, about $16/sq ft in all. Sam obtained bank financing on the job. It took about 3,000 man-hours to build the Felts home, about half of them logged by Sam himself.

Sam, who had never built a structure before, says, "This has been the most rewarding experience of my life. I feel that I have created a fun place in which to live. At least I'll do my part in conservation of energy." Fun. Rewarding. Energy conservation. These three terms seem to be used most often by people describing their cordwood experience.

Year of main construction work: 1979

The Chiasson Home

(Figures 109 and E1)

Jean-Luc Chiasson, of Robertville, New Brunswick, built his beautiful gambrel-roofed home in 1978. The house is a combination of the post-and-beam style and log-ends used as load-supporting. It has a full basement. The corner posts are 10-by-10's, but the walls are 16" thick and of cedar. The cordwood was dried just three months and Jean-Luc says, "We are caulking it with insulation wool, little by little." He adds that his mortar mix of 20 sand, 5 Portland, 8 lime did not shrink.

The materials for the house cost $15,000, and $2,000 was spent on labor. Like the Currans, Jean-Luc obtained a bank loan and built his 1600 sq ft home for about $10/sq ft. And this does not include the usable space in the basement. Jean-Luc's "pièce de résistance" with regard to cost is: "The reason the cost of the house is quite high is that the foundation and good windows cost us $9,000, but I don't regret the windows." Quite *high*, Jean-Luc?

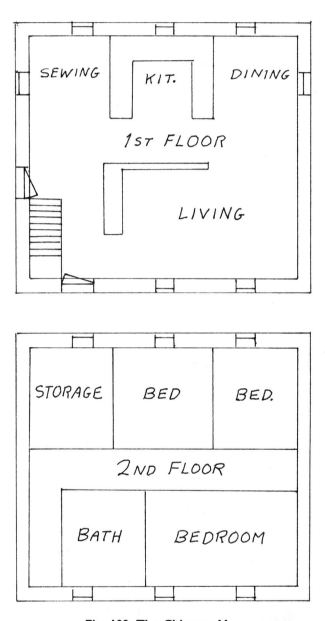

Fig. 109. The Chiasson Home.

Jean-Luc estimates that he put 1200 man-hours into the project and that other people also put in a total of about 1200 hours. He says that if he were to build the house again, he would not stack cordwood so high on the second story: "Stacking at the second story is slow and a lot harder, especially at a height of 25′ or so." Also, he would use his 4-by-4 window-framing vertically instead of horizontally.

The Bourgeois Addition

(Figures 110 and B1)

Euclide Bourgeois added a 17'-by-24', story-and-a-half shed to his house in Cocagne, New Brunswick, in 1974. He uses the building for wood storage and as a workshop. According to Euclide's brother, Leopold, the 12″ poplar cordwood was dried one year, but was not protected from the rain and was laying on the ground. The wood was not barked. It is no surprise to learn that the wood has shrunk considerably, but nothing is being done about it as the building is used primarily as a shed. Total man-hours on the project: about 500. Cost: $1,000 in 1974, or about $1.50/sq ft. Leopold says that they probably should have used squared timbers in the built-up corners and recommends that the bark should be removed from cordwood that is to be used in a dwelling.

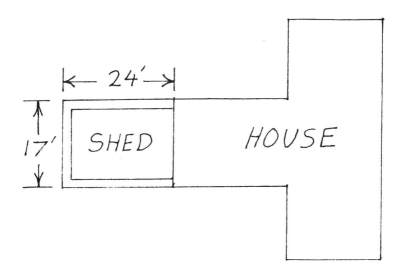

Fig. 110. The Bourgeois Addition.

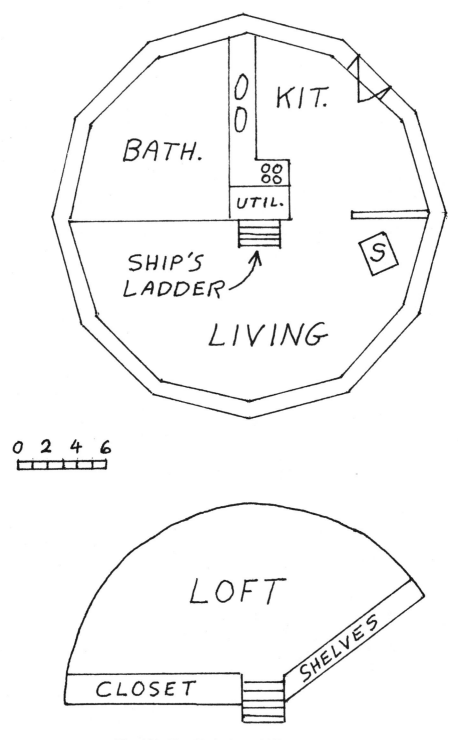

BATH.

KIT.

UTIL.

SHIP'S
LADDER

LIVING

S

0 2 4 6

LOFT

CLOSET

SHELVES

Fig. III. The Kwiatkowski Dome.

The Kwiatkowski Dome

(Figures 111, F1, and F2)

Tom Kwiatkowski visited our homestead three or four times prior to building his home; then I kind of lost track of him. In the summer of 1979 I heard that someone was building a cordwood house about ten miles from us, on Lake Champlain's Point-au-Roche. I followed directions to the site and found a 2-by-6 geodesic dome framework sitting on a beautiful 5'-high cordwood *kicker wall.* I noticed that the built-up corners were similar to a suggestion I made in an article appearing in *Farmstead #18* (Figure 68), and that the mortar mix looked a lot like the one I'd used at the Cave. But there was no one on site. I left a note, commenting on the excellence of the construction, and a few weeks later Tom and Helen Kwiatkowski and I finally touched bases. Tom's a fireman and works odd hours that are good for building but make him a hard guy to find sometimes.

The texture and relief of Tom's 12"-thick "firewood" walls is unique. Many of the log-ends come from a very old pile of firewood and all of them have dried at least two years. His walls are about ⅔ mixed hardwood, ⅓ softwood. His mortar mix was indeed the one I'd used at the Cave: 3 sand, 4 sawdust, 1 Portland, 1 lime, except that he was inclined to use more heaping shovels of sawdust than the other ingredients. His mix, then, is extremely high in sawdust content, so much so that the mortar joints have a slightly brownish tint. Yet the mortar remains hard and strong. Needless to say, his walls have experienced no mortar shrinkage and very little cordwood shrinkage. He used beige latex caulking to close up the few places where a crack had opened between wood and mortar. My understanding of wood shrinkage is that hardwoods generally shrink a lot less than softwoods, so this is one advantage of using hardwoods that somewhat offsets their lower R-factor.

The 700 sq ft story-and-a-half building cost about $8,500 to build, just over $12/sq ft. (Smaller homes usually cost more per square foot than larger homes — this is true for all types of construction.) Tom reckons he has 1,000 man-hours invested in the project and that 300 to 400 man-hours were put in by his new bride Helen and friends interested in the project.

"The house is proving to be low in cost to operate," says Tom. "I feel that this is due to its size, curvilinear shape, and the large amount of mass in the cordwood walls and interior — that is, brick floors and stonework. Also, the open interior design allows for excellent air circulation. We are heating with a Garrison II woodstove only, and will cut our wood down to a minimum with the addition of large triangular skylights this year.* I would recommend a floating slab for a foundation to anyone. With no excavation necessary, it is possible to do this job by yourself. Most important of all, though, is that the structure is an aesthetic, comfortable place in which to live."

* Correct placement of double-pane skylights on the south side of the dome will admit more heat by passive solar energy than they will lose at night, if shuttering or insulated drapes are used.

Log End Cottage

(Figures 112–114, and G2)

As Log End Cottage and Log End Cave are often referred to throughout this book, I will limit their "histories" to the basic facts.

Jaki and I built the post-and-beam Cottage in 1975. Our previous building experience was the construction of a 12' by 16' shed in which we lived during the house construction, although I had worked as a mason's laborer for a few months in Scotland. Our log-ends are 9" white cedar, dried for one year before laying up. Most of the walls are built with the rather confusing mix of 12 sand, 1 Portland, 1½ masonry, 1 lime. This mix was not a success. Shrinkage and mortar joint cracks resulted. An experimental panel of 6 sand, 9 sawdust, 1 Portland, 1½ masonry, and 2 lime did not shrink. We have since simplified this mix (as I've indicated elsewhere in the book). Complicated mixes invite error.

The house has about 500 sq ft of space in the basement, 500 sq ft on the first floor, and about 250 usable sq ft in the lofts. We spent about $5,000 on materials and $1,000 on contracting and labor. Half of the total went towards the basement and septic system, but even so, the house only cost $8/sq ft of actual living space in 1975. We could not have done as well without scrounging and recycling a lot of materials. If I were to build this design again, I would go for a floating slab instead of the basement, and I would build my lofts 2' higher to provide more usable space upstairs. Incidentally, the actual cost of the cordwood infilling, including insulation, mortar and saw rental, was only $112, easily the most economical part of construction.

Fig. 112. Log End Cottage.

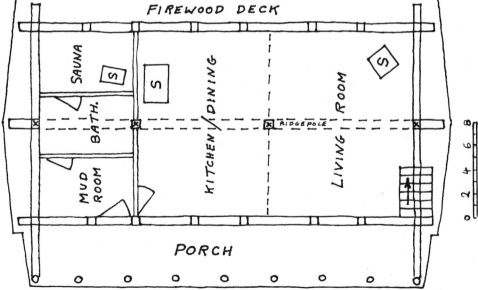

Fig. 113. Log End Cottage.

Fig. 114. Log End Cottage.

Log End Cave

(Figures 115–117, G1, and H2)

This earth-sheltered house features 10″ cedar cordwood as a low-cost means of infilling between the posts and beams of the south wall and the north gable end. Also, short (5″) cordwood serves as a means of closing in the space between rafters where they are supported by the main beams (Figure 106). After 2½ years at Log End Cottage, Jaki and I did not relish the idea of living in a house without at least a few of these weird and wonderful cordwood masonry panels.

The log-ends were force-dried in a sauna and next to the work stove during construction. Much of the cedar had been cut to 10′ lengths and barked two or three years earlier. Wood shrinkage has been minimal in this house and in the Cottage. I intend to apply a coat of thick white Thoroseal™ masonry sealer to the mortar joints of both houses when the job floats to the top of my priority list. I expect that this will fill the space between wood and mortar and cover many of the hairline mortar cracks at the Cottage. In the meantime, the slight wood shrinkage is not a problem. Occasionally, we stuff in a little insulation when we see daylight through an expanded check.

The mortar at the Cave (6 sand, 8 sawdust, 3 masonry, 1 lime) has not shrunk. I have since experimented with a mix reducing the sawdust content to six. This gives a slightly harder and smoother joint. Again, there is no shrinkage. Sam Felts, of Adel, Georgia, reports no mortar shrinkage when even less sawdust was used (see page 115).

The detailed cost analysis of our 910 sq ft house (1,036 sq ft gross) is reprinted from my earlier book, *Underground Houses* (page 125). Since this chart was compiled, we have spent about $350 more to finish off the place, or about $7,760 total for the basic house. This works out to about $8.50/usable sq ft of living space in 1976 prices. Approximately 2,000 man-hours were required to build Log End Cave, 1,500 of my own, 500 of others; some of it paid, some volunteered.

The house exceeds expectations. It is light, bright, airy, easy to live in, and phenomenally easy to heat. Our firewood requirement, mostly burned in an Oval air-tight cookstove (which also supplies our hot water) is about three full cords per year. We use bottled gas for refrigeration and summertime cooking and get our electricity for lighting, T.V., water pump, et cetera, from a Sencenbaugh 500-watt windplant.

Year of main construction work: 1978

Fig. 115. Log End Cave.

Fig. 116. Log End Cave.

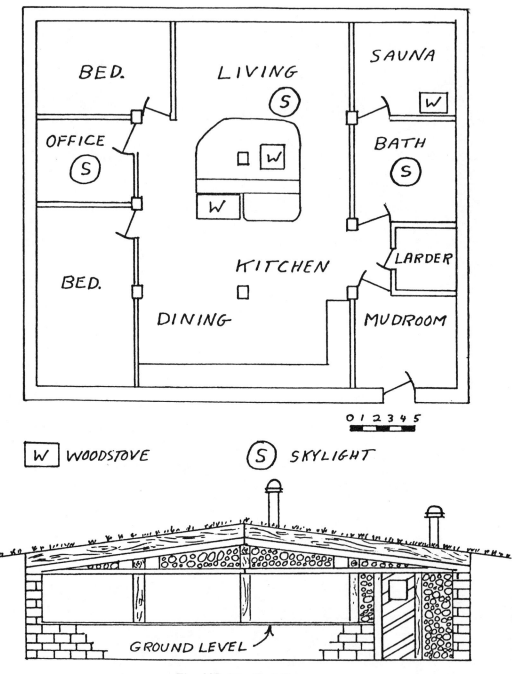

Fig. 117. Log End Cave.

Log End Cave—Cost Analysis

Heavy equipment contracting $892.00
Concrete 873.68
Surface bonding 349.32
Concrete blocks 514.26
Cement 47.79
Hemlock 345.00
Milling and planing 240.26
Barn beams 123.00
Other wood 167.56
Sheetrock 72.00
Particle board 182.20
Nails 62.88
Sand and crushed stone 148.21
Topsoil 295.00
Hay, grass seed, fertilizer 43.50
Plumbing parts 124.95
Various drain pipes 166.23
Water pipe 61.59
Metalbestos stovepipe 184.45
Styrofoam insulation 254.93
Roofing cement 293.83
Six mil black polyethylene 64.20
Flashing 27.56
Skylights 361.13
Thermopane windows 322.50
Interior doors and hardware 162.80
Tools, tool repair, tool rental 159.43
Miscellaneous 210.31
                                                              _____
Materials and contracting cost of house, landscaping and drainage $6,750.57
Labor 660.00
                                                              _____
Cost of basic house $7,410.57
Floor covering (carpets, vinyl, etc.) 309.89
Fixtures and appliances 507.00
                                                              _____
Total spending at Log End Cave $8,227.46

Fig. 118. Jack's "Ship with Wings."

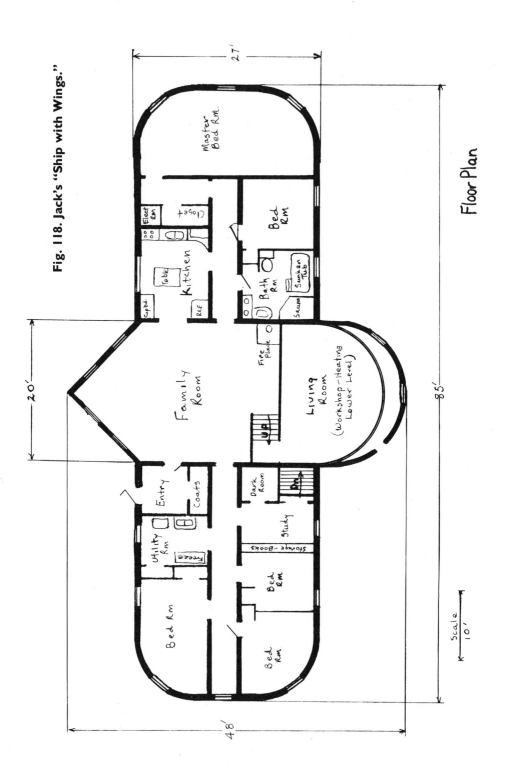

Floor Plan

Jack's "Ship with Wings"

(Figures 118, D1, and D2)
Year of main construction work: 1974

Jack and Helen Henstridge's 150-year-old frame house burned down in 1973. Compounding their problems was the fact that Jack had just lost a good job as a company pilot because of a business recession. They were truly up against it. Their assets were: a good piece of land on New Brunswick's St. John's River, $8,000 in fire insurance money, and plenty of kids and friends to help with building a house. The family moved into an old schoolhouse and Jack began to plan his new home. The end result was *The Ship with Wings,* probably the best known of all cordwood homes, having appeared in several American and Canadian periodicals, including *The Mother Earth News.*

Jack's 2,600 sq ft home features an elevated goedesic dome living room overlooking the river, and an airplane hanging from the gambrel ceiling. "We hung the airplane up as a sort of a joke," says Jack, "but then we decided that it was an ideal place to store it. What a mobile!"

The cordwood was about 80 percent softwood and 20 percent hardwood. Jack was in real need of shelter and wasn't too fussy about the kind of wood he used. The 9″ log-ends were laid up green, and considerable shrinkage resulted. Jack started stuffing the gaps with oakum, but has recently discovered that damp newspaper works much better. Jack spent about $6,000 on materials for his house and another $500 for electrical contracting, which comes to $2.50/sq ft. The east wing was never completely finished and is used mostly for storage.

Jack describes his home as "an experiment that worked exceedingly well and proves that it is quite possible to build with green wood and a 9″ wall. I couldn't have spent much less unless I used a clay mortar in the walls. The house is much larger than necessary, but the method is proved out. Green wood requires more 'afterwork' and the 9″ wall, while adequate, is just too hard to stack up. Sixteen-inch walls would be ideal."

All the trials and tribulations of building the *Ship with Wings* are detailed in Jack's entertaining book, *Building the Cordwood Home,* available from: Jack Henstridge, Box H, Upper Gagetown, N.B. E2V 2G2.

Had the late Canadian balladeer Robert Service met the venerable Henstridge, he might have been inspired to write:

The Ballad of Cordwood Jack

It was twenty below that fateful night the Henstridge house burned down,
The flames you could see a mile away, lighting Upper Gagetown.
"I felt like a fool, but went back to school for a place to live," says Jack,
"The insurance was paid, but it wasn't much, and I started to plan my shack."
The plan was for logs, but through swamp and bogs, he couldn't haul 'em out.
"We'll hafta cut 'em short," said Jack. His wife then started to shout.
"You can't do that! They'll be too small!" she earnestly reported.
"We'll lay 'em crosswise," said Jack. "We'll stack 'em all up mortared!"

Survival Shelter

(Figures 119 and E2)

On the way back from Jack's place a couple of years ago, Jaki and I stopped to look at Gottlieb Selte's small cordwood structure, which does double duty as a meditation *zendo* and a temporary shelter. This simple pentagonal building is an example of what can be done when one has to supply oneself with low-cost shelter. The house was built with dead wood — mostly budworm-killed spruce — has a sod roof, and cost $180 to build. The total time spent on the project, from on-site arrival until completion, was three weeks. Only hand-powered tools were used; even the water was carried in buckets for ⅛ mile. Gott's mortar mix of 7 sand, 2 sawdust, 1 masonry, 1 lime did not shrink, nor did the wood. The shelter is just temporary, but its free-form style is simple, warm, compact, and cheerful. The unusual shape of the house is the result of the builder trying to make maximum use of some recycled timbers as a beam foundation. Gott lives in Germany, but plans on returning to build a large cordwood house already started on the same site.

Thanks to Jack for helping with this article. As Gott and his friends place a high value on peace and quiet, we think it best not to give the location of this structure.

Year of main construction work: 1977

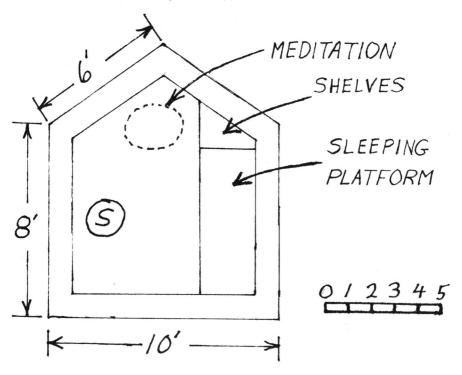

Fig. 119. Survival Shelter.

The Boyd Home

(Figures 120 and H3)

The largest circular cordwood house of which I am aware was built in Arkansas by Andy Boyd. The place is 50′ in diameter and has walls 18″ thick, a veritable fortress. The home has 1,963 sq ft gross, but 228 sq ft are lost to the cordwood walls, leaving a usable 1,735 sq ft. The cordwood is mixed oak, cedar, elm, pine, locust, and dogwood. Andy's mortar mix was 20 sand, 5 lime, 3 Portland. Both the wood and the mortar shrank. Andy tried various methods of chinking the resulting gaps, including latex and silicone caulking. His evaluation is that "slurry of mortar," an extremely thin mortar mix, is best.

Andy is very happy with his home and says that the round shape is the most positive thing about it. He heats his home with an attached solar greenhouse and has a Mother woodburning stove as a backup. The bathroom is at the core of the house and is skylighted.

One day Andy laid up about 30 sq ft of cordwood masonry, and work was interrupted by an Arkansas "frog strangler" — it rained for three days. The log-ends in Andy's new wall section swelled and broke the wall. This is why we recommend covering the work during construction and protecting the finished wall with a good roof overhang, although I have never heard of any problems from wood expansion on a finished wall.

The Boyds spent about $12,000 for materials in their house, and Andy, who had no previous building experience, built the house mostly by himself in about 2,000 hours. Others put in about 700 hours, including $1,500 of paid help. All told, then, the home cost about $13,500, or about $7/sq ft.

Year of main construction work: 1979

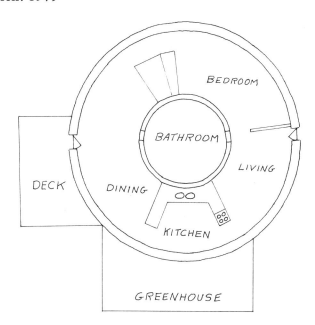

Fig. 120. The Boyd Home.

Ferny Richard's Cabin

(Figures 121 and B2)

Ferny Richard built his country hideaway near Rogersville, New Brunswick, for $1,000 in materials and $75 in hired help. All told, about 500 man-hours were spent on the project, half of them by Ferny. This points out another positive feature of building out of cordwood. Unlike the story of Tom Sawyer and the whitewashing, no con job is required to enlist help; people like to pitch in because it really is fun!

Ferny's cabin has about 395 sq ft of gross area, but the 12″ cordwood walls comprise about 75 sq ft, leaving 320 sq ft of usable floor area, still less than $3.50/sq ft. And Ferny says that now, with the experience under his belt, he could build a second, similar house for half the money!

The cordwood in this curved-wall cabin came from 40 fir logs that had been drying for 18 months, two old telephone poles, and a cord of very dry cedar. The fir shrank, but not the old poles and the cedar. Ferny says, "I built the cabin right on the ground. The first course of cordwood is old railroad ties, 18″ long. On top of this first course, I laid the floor joists and built them right into the wall."

Year of main construction work: 1975

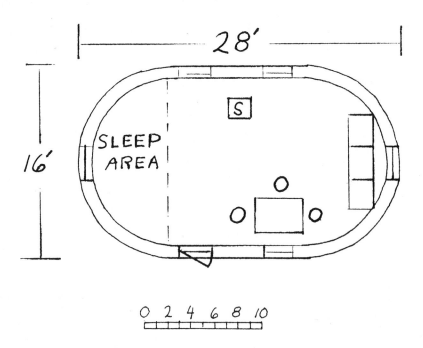

Fig. 121. Ferny Richard's Cabin.

La Maison de "Bois Cordé"

(Figure 122)

François Tanguay included some unique innovations in his round cordwood house near Inverness, Quebec. The first was his cellar foundation, made of 24″ cedar cordwood with the exterior covered with a thick black foundation coat. The backfill is sand, and a footing drain draws water away from the wall. François reports no water penetration into the cellar, which he uses for storage and for growing mushrooms. François expects his foundation to last for a long, long time. While I cannot recommend the placement of cordwood up against the earth in this fashion, it would be wrong to say that it can't be done, as François has shown.

The next two floors are fairly conventional; that is, of similar construction to the other round houses described in this chapter. The cordwood is 12″ cedar, dried for four years. The mortar mix was 5 sand, 5 sawdust, 2 Portland, 2 lime, quite similar to my own sawdust recipe, but a little stronger. The walls have experienced "very little cement cracking and hardly any wood shrinkage." There is "some wind penetration in the joints" and François expects that a whitewashing of the exterior will leach into the cracks and cure the problem.

Now listen to this: the Tanguay roof has the shape of a martini glass, with all the rain water flowing to a drain in the center of the roof. The soft water is carried by a pipe right down the central stem of the martini glass (post) and into a collecting tank in the basement.

François has about 600 usable sq ft per floor, or 1,200 in all not counting the basement. The total cost of $6,000, then, converts to about $5/sq ft. François and his friends put about 1,200 man-hours in "the basic house plus some finishing."

François told of his cordwood experience in *La Maison de "Bois Cordé"* (L'Aurore, Montreal, 1979), which includes a French translation of Jack's book.

Year of main construction work: 1978

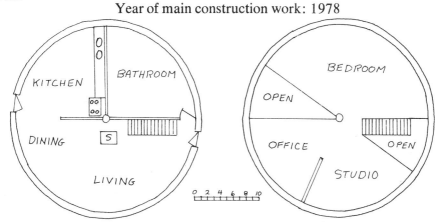

FIRST FLOOR   **Fig. 122. La Maison de "Bois Cordé."** SECOND FLOOR

Cliff Shockey's First Home

(Figure 123)

There is a shortage of trees in southern Saskatchewan, but that didn't stop Cliff Shockey from building the best-insulated cordwood home in this book. Cliff bought six miles of old cedar telephone poles that were scheduled to be replaced. This was his first smart move, as there was no wood shrinkage except when he used wood from near the base of the poles. His second good move was to lay a double wall of cordwood; that is, an 8″ outer wall, a 6″ inner wall, and 10″ of fiberglass insulation in-between. His outer walls join together with built-up corners; the inner walls meet at 6-by-6 corner posts. His house, then, is a hybrid of two styles. Cliff installed an electric heating system, but has yet to turn it on. The Shockeys heat through the severe Saskatchewan winter on a couple of cords of low-grade firewood, supplemented on sunny days by the solar gain through the large south-facing thermopane windows.

Cliff spent $8,700 on his 600 sq ft home, about $14.50/sq ft. The Shockeys' second home will be almost twice as big as the first. "The second house will cost me a lot more, probably $20,000," says Cliff. "It's the one we plan to spend the rest of our lives in, so we are finishing it a lot more fancy — like spiral staircase, whirlpool bath, attached greenhouse, outside shutters for the large south-facing windows, solar collector for hot water, et cetera." For all that, Cliff is paying only $17/sq ft in 1980 prices, and, as he's beefed up his roof insulation and is adding insulated shutters, I dare say the Shockeys will find their second house as easy to heat as the first.

Cliff says, "The satisfaction of designing and building your own home cannot be measured. It is truly one of the most rewarding things I have done. It's a lot of work, though. I spent eight months on the first house. People should be warned about all the labor that is involved in stackwall construction."

Cliff's right, of course. Cordwood masonry *is* labor intensive, although his super-insulated double-wall method is bound to add a lot of time to an already time-consuming job. Another idea along the same lines might be worth considering. The inner wall could be surface-bonded concrete blocks, with a white, textured finish. Fieldstones or very dry hardwood log-ends could be randomly set into this wall as design features, even shelves, perhaps, for holding candles or displaying keepsakes. Some of the advantages of this technique would be increased speed of construction, a brighter internal wall, greater mass-heat-storage ability, and an internal atmosphere which is bound to be less arid than in an all-cordwood home. One of the problems with cordwood is that the inside air can get extremely dry. We experienced 20 percent relative humidity at Log End Cottage in the winter when the woodstoves were going, which is rather low for a home. Humidifiers and open water containers on the stoves have little effect, as the log-ends just soak up an incredible

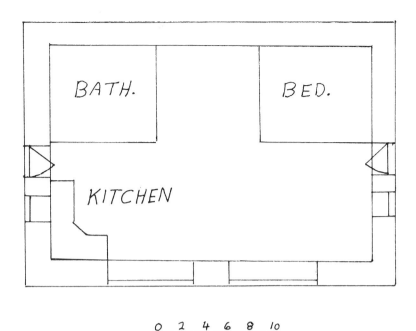

0 2 4 6 8 10

Fig. 123. Cliff Shockey's First Home.

amount of moisture. Cliff uses a polyethylene vapor barrier between his fiberglass insulation and the inner wall, which may help retain a little more humidity.

Someone once suggested to me that a house with stone on the exterior and cordwood on the interior would be a good combination. The idea was that the house would have the long life of stone and the warmth of the cordwood. The principles of heat storage and insulation and the proper placement of these substances in relationship to each other do not support this line of reasoning. The best use of the same material would be to put the cordwood on the outside and the stone on the inside. At first thought this seems a little disturbing. The stone on the inside has a "cold" connotation. At the same surface temperature, stone feels colder to the touch than cordwood, for example, because it has a much greater conductivity. Another problem with stone on the interior is that it soaks up a lot of light. This is why I suggest the blocks with a white textured surface applied to the interior. Of course, this double-wall method increases the cost of building, as well as the labor. There's no such thing as a free lunch.

Year of main construction work: 1977

The Gallant/Pond Home

(Figures 124 and A2)

One of the most beautiful cordwood homes I have seen was built by Auguste Gallant and Bonny Pond of Petit Rocher, New Brunswick. The house combines the post-and-beam and curved-wall styles. The cordwood is cedar, bought for about $2 per cord from the government (in their rush to get at the "good" wood multinational companies, cutting on Crown land, left the cedar to rot). The log-ends were only two or three months dried at their 12″ length, so there was considerable wood shrinkage. The owners are caulking with oakum to seal off air leaks, but they say, "This is a long, l–o–n–g process."

Auguste and Bonny used a mortar mix of 19 sand, 5 lime, and 3 Portland cement. They report no mortar shrinkage. Although the home is large — about 1,700 sq ft, according to the plans — Bonny took the time to incorporate a very artistic pointing technique, which she calls "routing." The mortar is recessed near the log-ends, but is allowed to protrude somewhat in-between. The effect is that both the cordwood and the mortar are featured in relief. This is similar to V-joint pointing used by stone masons. Bonny says, "It lengthens your stacking time, but the effect is worth it."

For comparison purposes, the materials cost of their mortgage-free home was $24,000. An additional $4,300 was spent on labor. The couple estimate that they could duplicate their experience for a little more than half of what they spent in 1978, when the bulk of the work was done. A more detailed cost analysis — including land, appliances, and a "Cadillac" heating system — shows that the total spending at the Gallant/Pond residence was $41,000:

Land and land preparation	$ 7,200
Foundation	5,700
Framing and exterior	3,700
Cordwood walls	1,200
Windows and roofing	6,000
Heating, plumbing, electricity	10,000
Greenhouse and patio	600
Interior framing and walls	1,000
Appliances	2,800
Tools and equipment	900
Miscellaneous	1,900
Total	$41,000 Canadian

Bonny says, "Our 'Cadillac' heating system consists of thermostatically controlled hot water baseboard heating with two separate furnaces. The main one is wood; the back-up is electric."

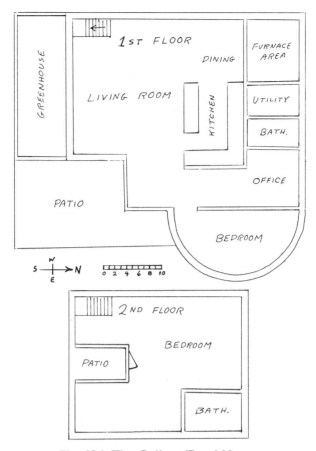

Fig. 124. The Gallant/Pond Home.

Auguste estimates that 212 man-days (10 hours per day) were required to build the basic shell. This does not include electrical, plumbing, heating, windows, and roofing, which were contracted out.

Bonny maintains that cordwood homes cannot be built by one person. "It is so time-consuming and, at times, frustrating, that the encouragement of a partner is a definite necessity." Her point is well taken. Cordwood construction benefits immensely from a team effort, especially with such a large house. A small home could be built by a single person with tenacity, but cordwood seems to draw volunteer labor so easily that I can't imagine the need to "go it alone."

All the hard work, the meticulous pointing, and the chinking must have been worth it to Auguste and Bonny, who proudly concluded a recent letter with these words:

"Our house is more than a beautiful, ecologically sound, economically attuned domicile. It is a statement of what we feel about ourselves, each other, and the world in which we live. Spread the word!"

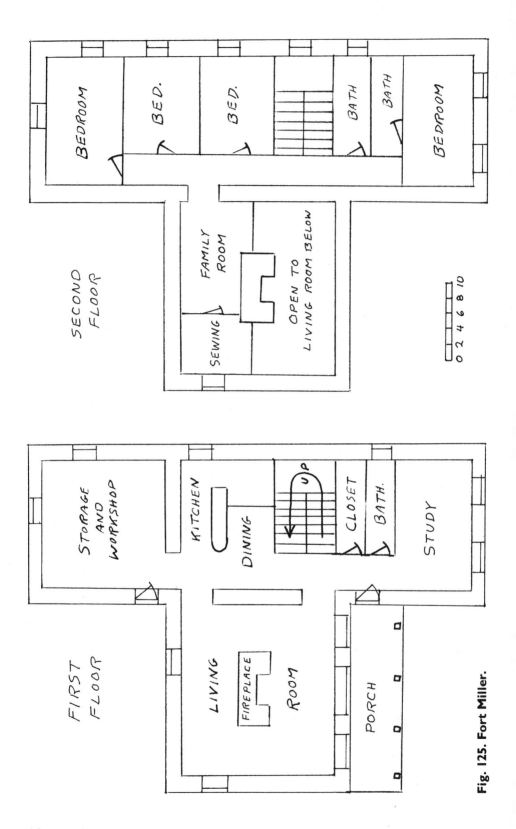

Fig. 125. Fort Miller.

Fort Miller

(Figures 125 and C1)

Other owners in the approved subdivision just outside of Fredericton, New Brunswick, were more than a little concerned when Malcolm Miller kept bringing cord after cord of cedar logs to his building lot. Their concern grew when chainsaws were running all day, day after day, cutting the random-length logs and a pile of 6-by-6 timbers into shorter lengths. Next, the logs were being stacked up with mortar as if the workers were trying to build a permanent woodpile . . . which was, of course, exactly what they were doing. Some of the people in the neighborhood got so concerned — to the point of fearing for their property values — that they thought a petition should be signed to put a stop to all this craziness. When Malcolm built his corners up to the 6′ height, and the 24″ cordwood walls up to 3′, some of the onlookers started to call the place "Fort Miller," because at this stage the huge house looked as much like a fort as anything else, turreted corners and all. As construction progressed, and the beauty of the house started to emerge, the petitions were torn up and people began to stop and ask about this obviously sound method of construction. When the house was finally closed in and the impressive architecture was evident to all, it was generally conceded that this crazy, bearded schoolteacher wasn't really so crazy after all.

The Millers ended up with a 3,700 sq ft house (3,100 usable) in an approved subdivision for a grand total of $73,000 Canadian, including carpets and all the trimmings. This works out to between $20 and $24/sq ft, depending on whether gross or net area is considered. Malcolm says that one doesn't get much built in the Fredericton area for less than $40/sq ft and that his house is probably worth close to $125,000, as of 1979. And he says that if he built a similar house again, with this experience under his belt and without all the darned codes and regulations, he could easily do it for $40,000. Things like an elaborate roof-framing system, three bathrooms, a very expensive electrical system run by conduit through the walls to meet tough codes, a huge stone masonry fireplace, expensive carpets, and a certain kind of roof shingle to meet planning regulations all took their toll on the final cost. Incidentally, Malcolm obtained a bank loan for the project, which is another step forward in giving cordwood some respectability.

Jack Henstridge, who lives fairly close to *Fort Miller,* took Jaki, her parents, and me over to the subdivision to see the house, which was then nearing completion. We were truly amazed by the power of the building. The house is massive, to be sure, but it also has very pleasing lines. An impressive architectural feature, which, unfortunately, cannot be appreciated from the road, is the large back wall with its five dormers projecting from the roof line. Malcolm's place reminds me of a manor house one would expect to find on an English country estate.

Malcolm found the experience extremely rewarding, though often trying. Not only is his a very large house — the kind of project that can destroy a man of lesser mettle — but Malcolm's experiences with the local building and planning authorities were often extremely frustrating. It was worth it, though, and Malcolm's experience is bound to make cordwood construction more acceptable for the next building department.

One particular construction problem that Malcolm reports is that the 24″ walls are so wide that it's hard for one man to lay both mortar joints, especially when the work is a few feet up. It's a much easier job with a worker on both sides of the wall.

Another observation of Malcolm's is well worth noting. He says that building up the corners a foot or two at a time is a nuisance and that it's hard to keep them plumb and square by the periodic return to this work. He suggests building a corner in one day by the method illustrated in Figures 126–127. All 6-by-6's used for these corners should have two holes drilled in them, one centered 3″ in from the end, and the other centered 18″ from the first. The holes should allow the insertion of a ½″ rebar. The first bed of 6-by-6's is laid in a 1″-thick bed of stiff mortar, with a few crushed stones — or round river pebbles — placed in the mortar bed to increase load-carrying ability until the mortar sets. Another method of spacing the timbers during mortar setting is to drive four 2″ roofing nails into the underside of each timber to within 1″ of home. The space between timbers should be infilled with log-ends as shown. Four 8′ rebars are inserted in the four holes, establishing the spacing for future courses. All succeeding mortar beds are laid similar to the first, with the little stones or nails used as spacers. All the rest of the 32″ timbers are threaded over the rebar as the corner is laid up, so that the whole corner is reinforced and incredibly strong. Even the plate could be tied into the corners later by placing the 2-by-6's over the top of the rebar.

Sounds good, Malcolm.

Year of main construction work: 1979

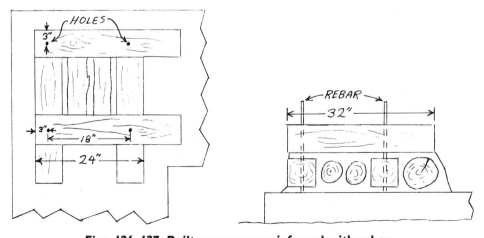

Figs. 126, 127. Built-up corners reinforced with rebar.

7. Special Effects

There are many good reasons to build with log-ends: wood unsuitable for other styles of building can be used; the work is not as physically demanding as for a "traditional" log cabin; the wide wall makes for excellent insulation and thermal mass; and the total cost of the house is low.

Still, one of the primary appeals of cordwood masonry is visual; somehow, this building style is faintly reminiscent of the gingerbread cottages in storybooks. Together with a sturdy post-and-beam framework, the look is similar to that of the English "black and white" houses, and, at the same time, has a certain Scandinavian flavor. The log-ends have such a natural beauty of their own that it would almost take a conscious effort to lay up a wall that is not pleasing to the eye. The different colored patterns of the end grain and the random checks that form during drying almost guarantee success. Almost.

But if time is taken, and attention to detail, improvisation, and, above all, imagination are exercised, your walls will be personal, unique, and very beautiful. I will present a few ideas in this chapter which I hope will illustrate that cordwood masonry can be a highly creative medium.

RELIEF

Recessed pointing by itself should ensure a three-dimensional look to the wall, but further interest is added by deliberately recessing a log-end or allowing one to protrude, especially if the log-end has some special color, texture, or natural design; or a log-end can be wood-burned with a hot wire and featured in relief. A good example of relief is shown in Figure 128.

VARIETY OF STYLE

When building Log End Cottage, we were able to obtain a couple of loads of scraps left over from cedar logs that had been used for traditional log

139

Fig. 128.
A good example
of relief.

cabins. The scraps ranged in length from 1′ to 3′ and were all uniformly milled to 6″. For variety, and to make use of this valuable windfall, we built three panels with the log-ends laid up horizontally in courses. The result is shown in Figure 129. By coincidence, the roof slope of one of the cabins was the same as that of our diagonals, and we only had to cut three or four pieces to fit. Some of the pieces were very wide, 12″ to 14″. These we laid together in a course to make a boot shelf for our mudroom.

Fig. 129. Two different styles of cordwood masonry.

PATTERN

We stayed with the random style throughout Log End Cottage, except for a few stained *beam-ends* that we placed in some of the square panels as focal points (see Figure 130). Another idea is to take a few log-ends of the same size and to feature them in a design. The drawings on page 142 illustrate some of the possibilities, which, except for the pentangle, take advantage of the natural hexagonal configuration which log-ends of the same size will assume. To accent the design, log-ends 2″ longer than usual should be used and the whole design should be allowed to extend 1″ proud of the rest of the wall, inside and out. The center log-end could be regular length, or it could be a "bottle-end" (see page 147).

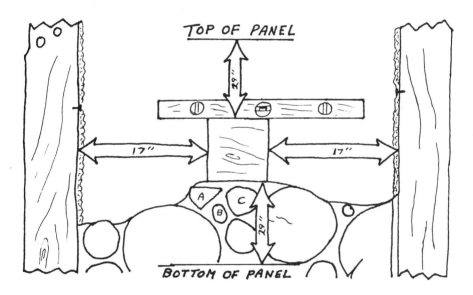

Fig. 130. Planning the placement of a "beam-end."

Stained beam-ends add interest to rectangular panels. They should be centered in the panel and leveled. Centering takes planning and measuring. Ends A, B, and C were used specifically to bring the work up to the right height for the beam-end.

HEXAGON   DIAMOND   TRIANGLE

STAR OF DAVID   PENTANGLE
(DIFFICULT)   (V. DIFFICULT)

Figs. 131, 132. Designs incorporated into panels of cordwood.

Fig. 133. This woodpile photographed near Scotia, New York would make a beautiful cordwood panel.

Design work of this kind takes advance planning — special log-ends should be put aside for the purpose — and it takes time, but it's worth it. Some of the best designs in the house, though, will be those that occur naturally in the random style. We have one panel, for example, where small log-ends seem to cascade over a precipice and smash on a large log-end below.

COLOR PRESERVATION

Cordwood loses its color and turns gray over the years, which is a pleasing effect of its own. An alternative is to dip the ends of the log-end ½″ into a pan of boiled linseed oil prior to laying up. This goes a long way towards preserving the end-grain color of the various woods, especially on the interior where discoloration will naturally be slight. Figure C2 shows an interior wall of split red pine which was preserved in this way. The visual effect is that of a wall built up of petrified wood.

While it is not strictly necessary to use a preservative on the exterior — if care is taken to keep the wood from staying in a damp condition — many builders have tried various preservatives. A preservative will probably help the weathered exterior retain some of its original color and is bound to help somewhat in prolonging the life of the wall, although it should have a long life anyway. The Northern Housing Committee recommends dipping the log-ends into "a 14% solution of copper sulfate (bluestone) and water for five minutes. If you are using a 55-gallon drum, dissolve 56 pounds of copper sulfate in approximately forty gallons of water."[10] If this dipping procedure is employed, the log-ends must be allowed to return to ambient moisture content before laying up. Otherwise, unacceptable wood shrinkage may result.

I have heard of one family using a penta preservative on the exterior, but I cannot advise the reader to follow suit. "Underground" builder and writer Mike Oehler says:

Penta (pentachlorophenol) does not entirely stay fixed in the treated wood. Much of it leaches out to move around in the environment. In 1976 researchers from the chemistry department tested students at Florida State University, Tallahassee, and discovered that, while 36% of the dormitory students had measurable levels of 2,4,5-T and Silvex (herbicides) in their urine, virtually *every* student tested showed traces of penta.

"Like many of the herbicides and some of the pesticides, the fungicide penta contains a contaminant created during the manufacturing process sometimes referred to as TCDD (2,3,7,8-tetra-chlorodibenzo-p-dioxin) but most commonly called dioxin. This is one of the most deadly chemicals known to man. The food and drug administration warns that it is "100,000 to a million times more potent" than thalidomide, the notorious drug which caused widespread birth defects in Europe.

"The use of penta has reportedly been placed under restriction by the Canadian government. It and all wood preservatives are now under review by the Environmental Protection Agency in the U.S. It is definitely recommended that penta *not* be used within a home.[11]

Incidentally, dry wood that has turned gray or black because of bacterial action on the wood sugars, can be returned to a rich brown color by scrubbing the end with a strong (15 percent) bleach solution. After a day's good drying, the end can be dipped in linseed oil to preserve the color.

SHELVES

Cordwood masonry walls lend themselves beautifully to the addition of shelves. At Log End Cottage we employed two different types of shelf. Figure 134 shows individual slat-ends protruding 7″ into the room. These make fine display shelves and are grand for candles. Another kind of shelf is also shown to the left of Figure 134, where a board crosses two protruding slat ends. *Use a carpenter's level to get this exactly right.* The shelf is placed to give continuity to the adjacent panels.

Fig. 134. Display shelves.

COAT PEGS AND AXE BRACKETS

These are little log-ends sticking out 4″ (Figure 135).

**Fig. 135.
Coat pegs.**

MAGAZINE RACK

While cutting firewood one day, I came across a maple branch which divided itself into five smaller branches. By trimming the "fingers," I was able to sculpt the branch into a fairly good human hand. Mortaring the piece into a large, naturally hollow, cedar log-end produced the unique magazine rack shown in Figure 136.

**Fig. 136.
Magazine rack.**

145

STAIRWAY

Our stairway to the loft looks as if it were made from log-ends. As a matter of fact, the ash steps do go right through the wall and form stairs on both sides, but the whole structure was built before any log-ends were laid in this panel. The steps are very heavy and the load is completely carried by the 3-by-10 runners. Fitting the log-ends around the stairs was exacting work. I advise laying up the easiest panels first to build up confidence and to improve technique before tackling this kind of panel.

Figure 137 shows the work in progress. The two upright log-ends awaiting placement are resting on the inside step of the second ash tread. The third stairstep is the only one of the six that is made of two separate (18″) pieces, unavoidable here because of the interference of the diagonal support members. The other five are a full 45″ long — 18″ on each side, with 9″ hidden in the wall. Figure 138 shows the finished stairs from the inside. The outer stairs serve as a fire escape from a "tea deck" that adjoins the loft.

Figs. 137, 138. (Left) Filling in the stairway panel at Log End Cottage. (Right) The finished stairs.

BOTTLE-ENDS

Jaki and I had seen beer bottle masonry in books, but we were concerned about heat loss and the accidental breaking of the necks after the bottles were laid up. So we used a bottle cutter to cut 4½″ tumblers from beer and wine bottles, and glued the pieces together with an epoxy resin glue. Then we sealed the joint with Mortite, a substance something like modelling clay and that is used for all sorts of sealing work. This gave us a kind of double-glazing. It is important that the bottles be kept dry inside when they are sealed. A week after laying up a panel on the south wall, we noticed that there was moisture in one of the bottle-ends. A few weeks later, that bottle-end cracked, the only one in the Cottage to do so. The cracked bottle-end is hardly noticeable and does not constitute a hazard, so we have left it alone.

Another way to make bottle-ends is by gluing two mustard or baby food jars together as shown in Figure 139.

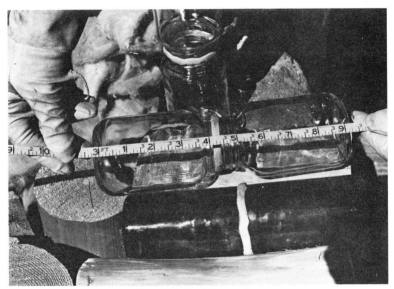

Fig. 139. Bottle-ends.

We used 36 bottle-ends at the Cottage, mostly as a feature of the panels next to the stairway, and 39 more in the 10″ cordwood walls at the Cave. I wish I could describe the jewel-like beauty of the direct rays of the sun playing on a panel of multicolored bottle-ends (see Figs. 140 and 141).

Bottle-ends cut and reglued as described would be useful in walls up to 12″ thick. For 16″ or 24″ walls, I advise laying up two uncut bottles with the uncapped necks pointing toward the wall's center. In this case, I feel that it is better not to try to seal the ends. The bottles should be able to breathe into the insulation cavity, releasing air pressure and preventing breakage. Care

147

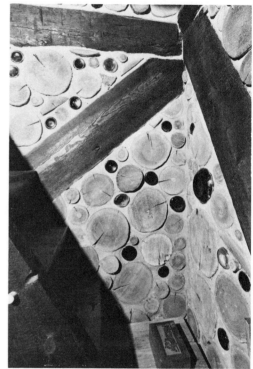

Figs. 140, 141. Bottle-ends by day . . . and by night.

should be taken to keep the insulation from hugging the bottle necks, for that would decrease the transfer of light through the walls. Inserting the neck ends of the bottles into a short cardboard cylinder might be useful for this purpose.

TERRARIUM

At the Cottage we laid up a one-gallon storage jar in one of the walls for use as a terrarium. Loosening or tightening the stained-glass cap regulates the air flow.

WINDOWS

Although our windows are framed with heavy timbers, cordwood masonry will easily support small windows randomly *floating* in a panel. Figures 142 and 143 show how such windows, framed with full-sized 2-by planking, could be built into a cordwood wall. Old windows, like the one in Figure 142, can often be found at auctions, garage sales, and building demolition sites. We had eight or ten such windows collected from here and there, but gave them to friends when we realized that the diagonals got in the way of using them in our own cottage.

The floating windows could be any shape, though curved frames might be difficult unless the builder is lucky enough to find portholes from a boat being broken up for salvage. I would dearly love to see a cordwood house with several randomly spaced portholes as a design feature. Another possibility is a framed stained-glass window floating in cordwood masonry, as shown in Figure 143. Again, galvanized roofing nails should be left projecting ½″ from the mortared side of all framing to help ensure a good wood-to-mortar bond.

Figs. 142, 143. Small windows, framed with full-size 2-by planking, can easily be built into a cordwood wall.

Imagination is the limit . . .
Happy stacking . . .
And may your shrinkage gaps be few and small!

ADDENDUM

The earthen roof waterproofing and insulation method described in this book was successfully employed at Log End Cave. However, I have since learned through discussions with several other builders and architects in the earth-sheltering field that the use of a hardboard layer to protect the rigid foam insulation is unnecessary, especially if extruded polystyrene (Dow Styrofoam®) is used. This material is very tough indeed and retains its R-factor underground, although basic caution should be exercised when walking on any unprotected foam insulation. Also, stones should be kept out of the first two inches of earth.

My thinking now is that the waterproofing should be applied directly on top of the roof planking and that the Styrofoam® should be placed immediately upon the membrane. This greatly simplifies construction and assures that the insulation is applied to the correct side of the waterproofing with regard to dew point. Although we have not had a problem with the Cave roof, it is safer to keep the membrane warm, guarding against condensation on its underside.

Based on discussions with experts in waterproofing, I am now of the opinion that neither the black plastic roofing cement and six-mil polythene method, nor the built-up roll roofing method would be correctly applied to wood decking. I would use either a $\frac{1}{16}''$ butyl membrane applied with an adhesive or the $\frac{1}{16}''$ Bituthene membrane which comes with its own adhesive characteristics. Both of these membranes have the ability to stretch over wood shrinkage gaps.

ROBERT L. ROY

APPENDIX

GLOSSARY

Bead of mortar—In cordwood masonry, one of the two narrow beds of mortar which enclose the insulation.

Bed—In masonry, the mortar upon which a brick, stone, or log-end is laid.

Berm—Earth banking.

Bottle-end—Two bottles cut and glued together to form a closed cylinder. Used like a log-end.

Buck—(verb) Cut into smaller pieces, as a log "bucked" into log-ends.

Checking—The natural splitting of a log-end (or any piece of wood) by rapid drying. Usually, only one major "check," or crack, will appear on each log-end.

Cordwood masonry—That style of building where short logs are laid up in a wall like firewood. The log-ends are bedded in a special mortar.

Feathering—To lap layer over layer so as to form a thin edge, known as a feather-edge.

Floated—In cordwood masonry, built in as a part of the wall with no other support except the mortar matrix.

Infilling—Masonry filling the space between major support pieces, such as posts and beams.

Kicker wall—A short supporting wall often used with geodesic dome construction to give additional height to the building.

Log-end—A short (6″ to 24″) log, one of many mortared together in a cordwood masonry wall.

Masonry cement—A mixture of Portland cement and lime, in a proportion of about two to one.

Mortar—Generally, a mixture of sand, cement, lime, and water, used for bedding and pointing bricks, stones, or log-ends. The characteristics of mortar depend upon the amount of each constituent.

Oakum—Rope-like caulking, impregnated with tar.

Panel—A section of cordwood masonry enclosed by posts and beams.

Percolation test—A test performed to quantify drainage characteristics of soil or earth, commonly used to determine drain field requirements for a septic system.

Plate—Planking used to cap a masonry wall. A plate distributes the roof load, can tie corners together, and provides a surface upon which to fasten rafters or floor joists.

Pointing—The filling of masonry joints with mortar, smoothed with a knife or the point of a trowel. Also called "grouting."

Proud—In masonry, the opposite of recessed. Protrusive.

Ridgepole, ridge beam—The major carrying beam of a roof system, supported by posts.

Screeding—Flattening concrete with the edge of a plank by vibrating it back and forth across the surface.

Slat-end—A 6″ to 24″ piece cut from the rough outside slat (or "slab") taken from a log being ripped into lumber.

Split-end—A split log-end.

Spud—A chisel-like tool for removing bark.

BIBLIOGRAPHY

Airhart, Sharon. 1976. Cord wood house. *Harrowsmith #4*: 54–7.

A century (or more) of stackwood homes. 1978. *The Mother Earth News #54*: 106–7.

Henstridge, Jack. 1978. Building the cordwood home. New Brunswick: Jack Henstridge.

Mann, Dale and Skinulis, Richard. 1979. The complete log house book. Toronto: McGraw-Hill Ryerson, Ltd.

Perrin, Richard W.E. 1974. Wisconsin's stovewood architecture. *Wisconsin Academy Review, Vol 20, #2:* 2–9.

The return of the cordwood house. 1977. *The Mother Earth News # 47:* 29–34.

Roy, Robert L. 1978. Building a log-end home. *Farmstead # 18:* 42–47.

Roy, Robert L. 1977. How to build log-end houses. New York: Sterling Pub. Co.

Roy, Robert L. 1979. Underground houses: how to build a low-cost home. New York: Sterling Pub. Co.

Square, David. 1978. Poor man's architecture. *Harrowsmith # 15:* 84–91.

Stackwall: how to build it. 1977. The Northern Housing Committee of the University of Manitoba.

Tishler, William H. Stovewood architecture. 1979. *Landscape Vol. 23 # 3:* 23–31.

SOURCE NOTES

[1] Square, David, "Poor Man's Architecture," *Harrowsmith #15* (1978), pp. 84–91.

[2] Henstridge, Jack, *Building the Cordwood Home,* St. Anne's Point Press, Box 691, Fredericton, New Brunswick (1977), p. 5.
This book is available from either the publisher, or from Jack Henstridge, Box H, Upper Gagetown, New Brunswick, Canada, E2V 262.

[3] Henstridge, Jack, *Building the Cordwood Home,* p. 35.

[4] Hodges, Tom, "The Peeling Spud: A Handy Tool for the Homestead," *The Mother Earth News* (July, 1976), p. 122.

[5] Haynes (Jr.), B. Carl and Simons, J.W., "Construction with Surface Bonding," Agriculture Information Bulletin No. 374, U.S. Dept. of Agriculture Research Service. This pamphlet is available from the Superintendent of Documents, U.S. Government Printing Office, Washington, D.C. 20402. Ask for Stock No. 0100–03340.

[6] Henstridge, Jack, *Building the Cordwood Home,* p. 39.

[7] *Stackwall: How to Build It,* The Northern Housing Committee of the University of Manitoba (1977), p. 17.

[8] *Stackwall: How to Build It,* p. 48.

[9] Henstridge, Jack, *Building the Cordwood Home,* p. 46.

[10] *Stackwall: How to Build It,* p. 16.

[11] Oehler, Mike, *The $50 and Up Underground House Book,* second edition, Mole Publishing Company, Bonners Ferry, Idaho (1979), p. 113.

PHOTO CREDITS

Color section photos A1, A2, B1, B2, C1, D1, D2, E1, E2, H1, by Jack Henstridge; C2 by Gene Lippis; F1, F2, H2 by Robert L. Roy; G1 by Pat Duniho; G2 by Dan Jerry; H3 by Andy Boyd. Black-and-white photos by Dan Jerry and Robert L. Roy.

ANOFAWM

23

Rohan Roy

Jack Heartridge

HAS BEEN FOUND QUALIFIED TO HOLD THIS EXALTED TITLE

MASTER MORTAR STUFFER

AND IS AN HONOURED MEMBER OF THE

ANCIENT and NOBLE ORDER of FREE THINKING and ACCREDITED

WOOD MASONS

IMHI

Jack Henstridge has designed and printed the certificate shown on the opposite page. For details on how to obtain a signed copy of the certificate, suitable for framing, write to Jack at Box H, Upper Gagetown, New Brunswick, Canada, E2V 2G2

This certificate, which is stamped with the seal of the Indigenous Materials Housing Institute, entitles you to put the initials "M.M.S." (Master Mortar Stuffer) after your name, which makes you just as fancy as anyone else. And the certificate, unlike many other lesser degrees and honors awarded, shows that you actually went out and did something to earn it, something good for you, your family, and the planet. Of course, the structure already proves that, but what the heck, everyone likes a little recognition.

ABOUT THE AUTHOR

Robert L. Roy left the American middle-class lifestyle in 1966 at the age of nineteen. During the next two years he traveled to over forty countries, finally settling in Scotland where he met Jacqueline Bates, now his wife. In 1974, the Roys visited the U.S. in search of land upon which to build their homestead. High land prices and restrictive building codes in Britain made homesteading there all but impossible.

In 1975, the Roys moved to West Chazy, New York, and built Log End Cottage, a cordwood structure. Rob wrote about the experience in *How to Build Log-End Houses* (Sterling, 1977). In 1977, Rob's new interest in earth-sheltered homes was manifested in the construction of Log End Cave and the publishing of *Underground Houses: How to Build a Low-Cost Home* (Sterling, 1979). Rob's articles on cordwood and earth-sheltered construction have appeared in *The Mother Earth News, Farmstead,* and *Alternative Sources of Energy.* In 1979, Rob co-founded Wood, Wind, and Earth, Inc., West Chazy, New York, which specializes in alternative energy and construction systems.

Rob and Jaki, and their young son, Rohan, live a self-sufficient lifestyle on their homestead, making electricity from windpower, heating with wood, and growing a significant part of their food.

Index